Learning with Partially Labeled and Interdependent Data

Massih-Reza Amini • Nicolas Usunier

Learning with Partially Labeled and Interdependent Data

Massih-Reza Amini
Laboratoire d'Informatique de Grenoble
Université Joseph Fourier
Grenoble
France

Nicolas Usunier
Université Technologique de Compiègne
Compiègne
France

ISBN 978-3-319-35390-6   ISBN 978-3-319-15726-9 (e-Book)
DOI 10.1007/978-3-319-15726-9

Springer Cham Heidelberg New York Dordrecht London
© Springer International Publishing Switzerland 2015
Softcover reprint of the hardcover 1st edition 2015

Printed on acid-free paper

Springer is part of Springer Science+Business Media (www.springer.com)

*All that is necessary, to reduce the whole
of nature to laws similar to those which
Newton discovered with the aid of the calculus,
is to have a sufficient number of observations
and a mathematics that is complex enough.*

Marquis Nicolas de Condorcet
1743–1794

Acknowledgements

The authors wish to thank the following those who made valuable suggestions or who have otherwise contributed to the preparation of the manuscript: Marianne Clausel, Cyril Goutte, François Laviolette, Clément Calauzènes, Patrick Gallinari and Guillaume Wisniewski.

Contents

List of Figures

Notation

$\mathcal{X} \subseteq \mathbb{R}^d$	Input space
$\mathcal{Y}$	Output space
$(\mathbf{x}, y) \in \mathcal{X} \times \mathcal{Y}$	An example $\mathbf{x}$ and its class y
$S = (\mathbf{x}_i, y_i)_{i=1}^m$	Labeled training set of size m
$X_{\mathcal{U}} = (\mathbf{x}_i)_{i=m+1}^{m+u}$	Unlabeled training set of size u
$\mathcal{D}$	Probability distribution generating the examples
$e : \mathcal{Y} \times \mathcal{Y} \to \mathbb{R}^+$	Instantaneous error function
$\mathcal{E}(f)$	Generalization error of function f
$E(f, S)$	Empirical error of function f on a training set S
$L(h, S)$	Convex surrogate function upper-bounding the empirical error on S
$R_u(f)$	Transductive risk of f estimated over $X_{\mathcal{U}}$
$m_{\varrho}(.)$	Unsigned margin of example $\mathbf{x}$
w	Weight vector of a learning function
$\mathcal{F}, \mathcal{G}, \mathcal{H}$	Classes of functions
$R(\mathcal{F}, S)$	Empirical Rademacher complexity of the class of functions $\mathcal{F}$ on a training set S
$\mathcal{R}_m(\mathcal{F})$	Rademacher complexity of the class of functions $\mathcal{F}$
$\sigma \in \{-1, 1\}$	Rademacher random variable; $P(\sigma = -1) = P(\sigma = +1) = \frac{1}{2}$
$\mathbb{H}$	An Hilbert space
$\Phi : \mathcal{X} \to \mathbb{H}$	A projection function
$K : \mathcal{X} \times \mathcal{X} \to \mathbb{R}$	A Kernel function
$\mathbb{1}_{[\pi]}$	Indicator function equal to 1 if the predicate π is true and 0 otherwise
$\langle ., . \rangle$	Dot product
$\mathfrak{T}$	Transformation function reducing a learning problem to the binary classification of pairs of examples
$\mathcal{R}_m^{\mathfrak{T}}(\mathcal{F})$	Fractional Rademacher complexity $\mathcal{F}$
$\chi(\mathcal{G})$	Chromatic number of $\mathcal{G}$

Chapter 1
Introduction

1.1 New Learning Frameworks

Consider the supervised learning task, where the prediction function, which infers a predicted output for a given input, is learned over a finite set of labeled training examples, where each instance of this set is a pair constituted of a vector characterizing an observation in a given vector space, and an associated *desired* response for that instance (also called desired output). After the training step, the function returned by the algorithm is sought to give predictions on new examples, which have not been used in the learning process, with the lowest probability of error. The underlying assumption in this case is that the examples are, in general, representative of the prediction problem on which the function will be applied. In practice, an error function measures the difference between the prediction of the model on an example and its desired output. From a given class of functions, the learning algorithm then chooses a function, which achieves the lowest empirical error on the examples of a given training set. This error is generally not representative of the performance of the algorithm on new examples. It is then necessary to have a second set of labeled examples, or a test set, and estimate the average error of the function and that will be this time representative of its generalization error. We expect that the learning algorithm produces a function that will have a good generalization performance and not the one that is able to perfectly reproduce the outputs associated to the training examples. Figure 1.1 illustrates this principle. Guarantees of learnability of this process were studied in the theory of machine learning largely initiated by Vapnik (1999). These guarantees are dependent on the size of the training set and the complexity of the class of functions where the algorithm searches for the prediction function.

Historically, the two main tasks developed under the supervised learning framework were classification and regression. These tasks are similar except over the definition of the desired outputs spaces; while in classification, the output space is

© Springer International Publishing Switzerland 2015
M.-R. Amini, N. Usunier, *Learning with Partially Labeled and Interdependent Data*,
DOI 10.1007/978-3-319-15726-9_1

Fig. 1.1 Illustration of training and test steps involved in supervised learning problems. In the learning step, a function minimising the empirical error over a training set is found from a class of predefined functions (schematically represented by *solid lines*). In the test step, outputs for new examples are predicted by the learned prediction function (*dashed lines*)

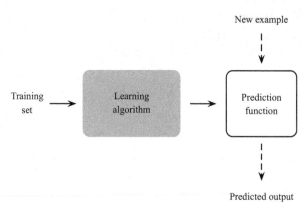

discrete (corresponding to a set of class labels), it is the real-value space in the case of regression.

Emerging technologies, particularly those related to the development of Internet, reshaped the domain of machine learning with new learning frameworks that have been studied to better tackle the related problems.

One of these frameworks concerns the problem of learning with partially labeled data, or semi-supervised learning, which development is motivated by the effort that has to be made to construct labeled training sets for some problems, while large amount of unlabeled data can be gathered easily for these problems. The inherent assumption, in this case, is that unlabeled data contain relevant information about the task that has to be solved, and that it is a natural idea to try to extract this information so as to provide the learning algorithm more evidence. From these facts were born a number of works that intended to use a small amount of labeled data simultaneously with a large amount of unlabeled data to learn a prediction function.

Another framework that attracted significant attention from the learning community since 2000 concerns the development of ranking models initially motivated by Information Retrieval problems and was later extended to more general issues.

For many years, learning algorithms developed following these frameworks have been successfully applied to a variety of problems including: recommendations systems, document classification, search engines, computer vision, etc.

1.2   Outline

This book presents the scientific foundations of the supervised learning theory as well as the semi-supervised and learning to rank frameworks mentioned above. The latter will pave the way to present the theory of learning with interdependent data which occurs when one tries to reduce some learning to rank tasks to binary classification of pairs of examples. This book is organized into three main chapters and the sequencing of ideas presented in each of them is as follows:

- In Chap. 2, we present an introduction to the statistical learning theory proposed by Vapnik (1999). We explain the concept of consistency of the empirical risk minimization principle which lead to the development of most supervised learning algorithms. The study of consistency paves the way to the presentation of the second principle of machine learning that is the structural risk minimization, opening the field to the development of new learning models. In particular, we present, how to derive generalization error bounds by describing the basic assumptions and the tools allowing to obtain them.
- We then outline the framework of semi-supervised learning in Chap. 3. we begin by describing the basic assumptions of semi-supervised learning and then depict the three main approaches developed based on these assumptions. In particular, we propose a bound on the transductive risk of a voted classifier that involves the margin distribution of the classifier and a risk bound on its associate Gibbs classifier. The bound is tight when the Gibbs's bound is so and when the errors of the majority vote classifier are concentrated on a region of low margin. As an application, we show how the bound can be used to automatically tune the threshold of a self-learning algorithm which iteratively assigns pseudo-labels to the set of unlabeled training examples that have their margin above the latter and extend the algorithm to the case of multi-view learning.
- In Chap. 4, we present a general framework for learning classifiers with non-iid data that occurs when reducing some learning problems to the binary classification of interdependent pairs of objects. We illustrate this situation by considering some cases of learning to rank as well as the multiclass classification task. Following a concentration inequality proposed by Janson (2004) for interdependent random variables, we propose a generic generalization error bound based on an extension of the Rademacher complexity, and show how this generic bound can be specialized to different cases of learning to rank.

Chapter 2
Introduction to Learning Theory

In logic, the process of finding a general rule from a finite set of observations is called induction (Genesereth and Nilsson 1987, Chap. 7, pp. 161–176)[1]. In machine learning, this inductive framework has been implemented according to the empirical risk minimization (ERM) principle, and its statistical properties have been studied in the theory developed by Vapnik (1999). The striking result of this theory is an upper bound of the generalization error of the learned function, expressed in terms of its empirical error on a giving training set and the complexity of the class of functions in use. This complexity reflects the ability or capacity of the function class to solve the prediction problem and is as larger as it contains functions able to predict all possible outputs of observations in the training set. In other terms, the higher the capacity, the lower the empirical risk and the less we are guaranteed to achieve the main goal of the learning that is to have a low generalization error. Hence, we see that there is a compromise between the empirical error and the capacity of the function class, and that the aim of learning is to minimize the empirical error while controlling the capacity of the function class. This principle is called structural risk minimization and along with the ERM principle they are at the origins of many learning algorithms. Further, they also explain the behaviours of algorithms designed before the establishment of the learning theory by Vapnik (1999). The remainder of this chapter is devoted to a more formal presentation of these concepts under the binary classification setting which constitutes the initial part of the development of this theory.

2.1 Empirical Risk Minimization

In this section, we present the ERM principle by first presenting the notations that will be further used.

---

[1] The opposite reasoning, called deduction, is based on axioms and consists in producing specific rules (which are always true) as consequences of the general law.

© Springer International Publishing Switzerland 2015
M.-R. Amini, N. Usunier, *Learning with Partially Labeled and Interdependent Data*,
DOI 10.1007/978-3-319-15726-9_2

2.1.1 Assumption and Definitions

We assume that the observations are represented in a vector space of fixed dimension d, $\mathcal{X} \subseteq \mathbb{R}^d$. The desired output of observations are assumed to belong to the set $\mathcal{Y} \subset \mathbb{R}$. Until the early 2000s, classification and regression were the two main frameworks developed in supervised learning. In classification, the output space $\mathcal{Y}$ is a discrete set and the prediction function $f : \mathcal{X} \rightarrow \mathcal{Y}$ is called a classifier. When $\mathcal{Y}$ is a continuous interval of the real set, f is said to be a regression function. In Chap. 4, we present a new framework called, learning to rank that has recently been developed in both machine learning and information retrieval communities. A pair $(\mathbf{x}, y) \in \mathcal{X} \times \mathcal{Y}$ thus denotes a labeled example and $S = (\mathbf{x}_i, y_i)_{i=1}^m \in (\mathcal{X} \times \mathcal{Y})^m$ indicates a training set. In the particular case of binary classification that we consider in this chapter, we note the output space by $\mathcal{Y} = \{-1, +1\}$ and a pair $(\mathbf{x}, +1)$ (respectively $(\mathbf{x}, -1)$) is called a positive (respectively negative) example.

The fundamental assumption of the learning theory is that examples are independently and identically distributed (i. i. d.) according to a fixed, but unknown, probability distribution $\mathcal{D}$. The identically distributed hypothesis ensures that observations are stationary while the independently distributed hypothesis states that each individual example provides a maximum information to solve the prediction problem. According to this hypothesis for a given prediction problem, examples of different training and test sets are all supposed to have the same behaviour.

Hence, this assumption characterizes the notion of representativeness of training and test sets with respect to the prediction problem, i.e. the training and unseen examples are assumed to be generated from a fixed source of information.

Another basic concept in machine learning is the notion of loss, also referred by risk or cost. For a given prediction function f, the disagreement between the desired output y of an example $\mathbf{x}$ and the prediction $f(\mathbf{x})$ is measured by an instantaneous loss function defined as:

$$e : \mathcal{Y} \times \mathcal{Y} \rightarrow \mathbb{R}^+$$

Generally this function is a distance in the output space $\mathcal{Y}$ measuring the difference between the desired and the predicted outputs for a given observation. In the regression framework, the usual instantaneous loss functions are the ℓ_1 and ℓ_2 norms of the difference between both the desired and the predicted outputs of an observation. In binary classification the most common instantaneous loss considered is the 0/1 loss, which for an exemple $(\mathbf{x}, y)$ and a prediction function f is defined as

$$e(f(\mathbf{x}), y) = \mathbb{1}_{[f(\mathbf{x}) \neq y]}$$

Where $\mathbb{1}_{[\pi]}$ is equal to 1 if the predicate π is true and 0 otherwise. In the case of binary classification, the learned function $h : \mathcal{X} \rightarrow \mathbb{R}$ is generally a real-value function and the associated classifier $f : \mathcal{X} \rightarrow \{-1, +1\}$ is defined by taking the signe over the output predictions provided by h. In this case, the equivalent instantaneous loss to the 0/1 loss, defined for the function h is

$$e_0 : \mathbb{R} \times \mathcal{Y} \to \mathbb{R}^+$$

$$(h(\mathbf{x}), y) \mapsto \mathbb{1}_{[y \times h(\mathbf{x}) \leq 0]}$$

Based on the definition of an instantaneous loss and the i. i. d. assumption over the generation of examples with respect to a probability distribution $\mathcal{D}$, the generalization error of a learned function $f \in \mathcal{F}$ is defined as:

$$\mathcal{E}(f) = \mathbb{E}_{(\mathbf{x},y) \sim \mathcal{D}} e(f(\mathbf{x}), y) = \int_{\mathcal{X} \times \mathcal{Y}} e(f(\mathbf{x}), y) d\mathcal{D}(\mathbf{x}, y) \qquad (2.1)$$

Where $\mathbb{E}_{(\mathbf{x},y) \sim \mathcal{D}} X(\mathbf{x}, y)$ is the expectation of the random variable X when $(\mathbf{x}, y)$ follows the probability distribution $\mathcal{D}$. As $\mathcal{D}$ is unknown, this generalization error cannot be estimated. In practice, the performance of a function f is estimated over a finite sample set S of size m using its empirical error (or empirical risk) defined as:

$$E(f, S) = \frac{1}{m} \sum_{i=1}^{m} e(f(\mathbf{x}_i), y_i) \qquad (2.2)$$

As examples are generated i. i. d. with respect to $\mathcal{D}$; it is easy to see that the empirical error is an unbiased estimator of the generalization error. Thus, in order to resolve the classification problem for which we have a training set S, it seems natural to choose a function class $\mathcal{F}$ and to search f_S within this class which minimises the empirical error over S.

2.1.2 The Statement of the ERM Principle

This principle is called empirical risk minimization (ERM), and it is the source of many algorithms in machine learning.

The fundamental question which arises here is *according to the ERM principle, is it possible to find a prediction function from a finite set of examples, which has a good generalization performance in all cases?* The answer to this question is, of course, no. For proof, consider the following toy problem

Example **Overfitting**
Suppose that the input dimension is $d = 1$, let the input space $\mathcal{X}$ be the interval $[a, b] \subset \mathbb{R}$ where a and b are real values such that $a < b$, and suppose that the output space is $\{-1, +1\}$. Moreover, suppose that the distribution $\mathcal{D}$ generating the examples $(\mathbf{x}, y)$ is an uniform distribution over $[a, b] \times \{-1\}$. In other terms, examples are randomly chosen over the interval $[a, b]$, and for each observation $\mathbf{x}$ its desired output is -1.

Consider now, a learning algorithm which minimizes the empirical risk by choosing a function in the function class $\mathcal{F} = \{f : [a, b] \to \{-1, +1\}\}$ (also denoted as $\mathcal{F} = \{-1, +1\}^{[a,b]}$) in the following way ; after reviewing a training set $S = \{(\mathbf{x}_1, y_1), \ldots, (\mathbf{x}_m, y_m)\}$ the algorithm outputs the prediction function f_S such that

$$f_S(\mathbf{x}) = \begin{cases} -1, & \text{if } \mathbf{x} \in \{\mathbf{x}_1, \ldots, \mathbf{x}_m\} \\ +1, & \text{otherwise} \end{cases}$$

In this case, the found classifier has an empirical risk equal to 0, and that for any given training set. However, as the classifier makes an error over the entire infinite set $[a, b]$ except on a finite training set (of zero measure), its generalization error is always equal to 1.

2.2 Consistency of the ERM Principle

The underlying question to the previous question is: *in which case the ERM principle is likely to generate a general learning rule?* The answer of this question lies in a statistical notion called consistency. This concept indicates two conditions that a learning algorithm has to fulfil, namely *(a)* the algorithm must return a prediction function whose empirical error reflects its generalization error when the size of the training set tends to infinity. Moreover, *(b)* in the asymptotic case, the algorithm must allow to find the function which minimises the generalization error in the considered function class. Formally:

$(a) \forall \epsilon > 0, \lim_{m \to \infty} \mathbb{P}(|E(f_S, S) - \mathcal{E}(f_S)| > \epsilon) = 0$, denoted as, $E(f_S, S) \xrightarrow{\mathbb{P}} \mathcal{E}(f_S)$;

$(b) E(f_S, S) \xrightarrow{\mathbb{P}} \inf_{g \in \mathcal{F}} \mathcal{E}(g)$.

These two conditions imply that the empirical error $E(f_S, S)$ of the prediction function found by the learning algorithm over a training S, f_S, converges in probability to its generalization error $\mathcal{E}(f_S)$ and $\inf_{g \in \mathcal{F}} \mathcal{E}(g)$ (Fig. 2.1).

A natural way to analyze the condition *(a)* of the consistency, expressing the concept of generalization, is to use the following inequality:

$$|\mathcal{E}(f_S) - E(f_S, S)| \leq \sup_{g \in \mathcal{F}} |\mathcal{E}(g) - E(g, S)| \tag{2.3}$$

We see from this inequality that a sufficient condition for generalization is that the empirical risk of prediction function, whose absolute difference between its empirical and generalization errors among all the functions in $\mathcal{F}$ is the largest, tends to the generalization error of the function, that is:

$$\sup_{g \in \mathcal{F}} |\mathcal{E}(g) - E(g, S)| \xrightarrow{\mathbb{P}} 0 \tag{2.4}$$

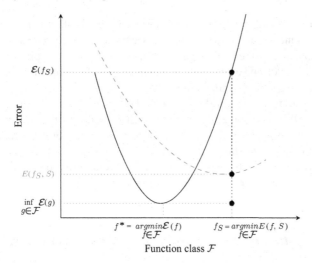

Fig. 2.1 Schematic description of the concept of consistency. The x-axis represents a function class $\mathcal{F}$, the empirical error (dotted line) and the generalization error (solid line) are shown as a function of $f \in \mathcal{F}$. The ERM principle consists in finding a function f_S in the function class $\mathcal{F}$ by minimizing the empirical error over a training set S. The principle is consistent, if $E(f_S, S)$ converges in probability to $\mathcal{E}(f_S)$ and $\inf_{g \in \mathcal{F}} \mathcal{E}(g)$

This sufficient condition for generalization is a consideration in worse case, and following Eq. 2.3, it implies a bilateral uniform convergence for all functions in the class $\mathcal{F}$. Moreover, the condition (2.4) does not depend to the considered algorithm but just to the function class $\mathcal{F}$. Thus, a necessary condition for the consistency of the ERM principle is that the function class should be restricted (see the overfitting example of the previous section).

The fundamental result of the learning theory (Vapnik 1999, theorem 2.1, p. 38) concerning the consistency of the ERM principle, exhibits another relation involving the supremum over the function class in the form of an unilateral uniform convergence and which stipulates that:

The ERM principle is consistent if and only if:

$$\forall \epsilon > 0, \lim_{m \to \infty} \mathbb{P}\left(\sup_{f \in \mathcal{F}} [\mathcal{E}(f) - E(f, S)] > \epsilon \right) = 0 \tag{2.5}$$

A direct implication of this result is a uniform bound over the generalization error of all prediction functions $f \in \mathcal{F}$ learned on a training set S of size m and which writes:

$$\forall \delta \in \,]0, 1], \mathbb{P}\left(\forall f \in \mathcal{F}, (\mathcal{E}(f) - E(f, S)) \leq \mathfrak{C}(\mathcal{F}, m, \delta) \right) \geq 1 - \delta \tag{2.6}$$

Where $\mathfrak{C}$ depends on the size of the function class, the size of the training set, and the desired precision $\delta \in \,]0, 1]$. There are different ways to measure the size of a function class and the measure commonly used is called complexity or the capacity of the function class. In this chapter, we present two such measures, namely the VC dimension and the Rademacher complexity) leading to different types of generalization bounds and also to the second principle of machine learning which is known as structural risk minimization (SRM).

Before presenting a generalization error bound estimated over a training set which is used to find the prediction function, we will consider in the first instance the estimation of the generalization error of a learned function over a test set (Langford 2005). The aim here is to show that it is possible to have an accurate upper-bound of the generalization error by using a test set that was not used to find the prediction function, and that independently of the capacity of the function class in use.

2.2.1   *Estimation of the Generalization Error Over a Test Set*

Remind that the examples of a test set are generated i. i. d. with respect to the same probability distribution $\mathcal{D}$ which has generated the training set. Consider f_S a learned function over the training set S, and let $T = \{(\mathbf{x}_i, y_i); i \in \{1, \ldots, n\}\}$ be a test set of size n. As the examples of this set are not seen in the training phase, the function f_S does not depend on the values of the instantaneous errors of examples $(\mathbf{x}_i, y_i)$ of this set, and the random variables $(f_S(\mathbf{x}_i), y_i) \mapsto e(f_S(\mathbf{x}_i), y_i)$ can be considered as the independent copies of the same random variable:

$$\mathbb{E}_{T \sim \mathcal{D}^n} E(f_S, T) = \frac{1}{n} \sum_{i=1}^{n} \mathbb{E}_{T \sim \mathcal{D}^n} e(f_S(\mathbf{x}_i), y_i)$$

$$= \frac{1}{n} \sum_{i=1}^{n} \mathbb{E}_{(\mathbf{x}, y) \sim \mathcal{D}} e(f_S(\mathbf{x}), y) = \mathcal{E}(f_S)$$

Thus, the empirical error of f_S on the test set, $E(f_S, T)$ is an unbiased estimator of its generalization error.

Furthermore, for each example $(\mathbf{x}_i, y_i)$ let X_i be the random variable $\frac{1}{n} e(f_S(\mathbf{x}_i), y_i)$, as all the random variables $X_i, i \in \{1, \ldots, n\}$ are independent and that they take values in $\{0, 1\}$; by noting that $E(f_S, T) = \sum_{i=1}^{n} X_i$ and $\mathcal{E}(f_S) = \mathbb{E}\left(\sum_{i=1}^{n} X_i\right)$, we have the following using Hoeffding (1963) inequality

$$\forall \epsilon > 0, \mathbb{P}\left([\mathcal{E}(f_S) - E(f_S, T)] > \epsilon\right) \leq e^{-2n\epsilon^2} \tag{2.7}$$

To better understand this result, let solve the equation $e^{-2n\epsilon^2} = \delta$ with respect to ϵ, hence we have $\epsilon = \sqrt{\frac{\ln 1/\delta}{2n}}$ and:

$$\forall \delta \in]0, 1], \mathbb{P}\left(\mathcal{E}(f_S) \leq E(f_S, T) + \sqrt{\frac{\ln 1/\delta}{2n}}\right) \geq 1 - \delta \tag{2.8}$$

For a small δ, according to Eq. 2.8 we have the following inequality which stands with high probability and all test sets of size n:

$$\mathcal{E}(f_S) \le E(f_S, T) + \sqrt{\frac{\ln 1/\delta}{2n}}$$

From this result, we have a bound over the generalization error of a learned function which can be estimated using any test set, and in the case where n is sufficiently large, this bound gives a very accurate estimated of the latter.

Example **Estimation of the generalization error over a test set (Langford 2005)**

Suppose that the empirical error of a prediction function f_S over a test set T of size $n = 1000$ is $E(f_S, T) = 0.23$. For $\delta = 0.01$, i.e. $\sqrt{\frac{\ln (1/\delta)}{2n}} \approx 0.047$, the generalization error of f_S is upperbounded by 0.277 with a probability at least 0.99.

2.2.2 A Uniform Generalization Error Bound

For a given prediction function, we know how to bound its generalization error by using a test set that did not serve to find the parameters of the latter. As part of the study of the consistency of the ERM principle, we would now establish a uniform bound on the generalization error of a learned function depending on its empirical error over a training set. We cannot reach this result, by using the same development than the one presented in the previous section.

This is mainly due to the fact that when the learned function f_S has knowledge of the training data $S = \{(\mathbf{x}_i, y_i); i \in \{1, \ldots, m\}\}$, random variables $X_i = \frac{1}{m} e(f_S(\mathbf{x}_i), y_i); i \in \{1, \ldots, m\}$ involved in the estimation of the empirical error of f_S on S, are all dependent to each other. Indeed, if we change an example of the training set, the selected function f_S will also change, as well as the instantaneous errors of all the other examples. Therefore, as the random variables X_i can not be considered independently distributed, we are no longer able to apply Hoeffding (1963) inequality.

In the following, we will outline a uniform bound on the generalization error by following the framework of Vapnik (1999). In the next section, we present another, more recent framework for establishing such bounds, developed in the early 2000s by showing the link with the work of Vapnik (1999). For the uniform bound, our starting point is the upper bounding of

$$\mathbb{P}\left(\sup_{f \in \mathcal{F}} [\mathcal{E}(f) - E(f, S)] > \epsilon \right)$$

At this point, there are two cases to consider, the case of finite and infinite sets of functions.

2.2.2.1   Case of Finite Sets of Functions

Consider a function set $\mathcal{F} = \{f_1, \ldots, f_p\}$ of size $p = |\mathcal{F}|$. The generalization bound consists in estimating the probability of $\max_{j \in \{1,\ldots,p\}} [\mathcal{E}(f_j) - E(f_j, S)]$ being higher than a fixed $\epsilon > 0$, for a given training set of size m.

If $p = 1$, the only choice of selecting the prediction function within $\mathcal{F} = \{f_1\}$ is to take f_1 and that before looking at the examples of any training set S of size m. In this case, we can directly apply the previous bound (Eq. 2.7) from Hoeffding (1963) inequality:

$$\forall \epsilon > 0, \mathbb{P}\left(\max_{j=1} \left[\mathcal{E}(f_j) - E(f_j, S)\right] > \epsilon\right) = \mathbb{P}\left([\mathcal{E}(f_1) - E(f_1, S)] > \epsilon\right) \leq e^{-2m\epsilon^2}$$

In the case where $p > 1$, we first note that

$$\max_{j \in \{1,\ldots,p\}} [\mathcal{E}(f_j) - E(f_j, S)] > \epsilon \Leftrightarrow \exists f \in \mathcal{F}, \mathcal{E}(f) - E(f, S) > \epsilon \qquad (2.9)$$

For a fixed $\epsilon > 0$ and for each function $f_j \in \mathcal{F}$; consider the set of samples of length m on which the generalization error of f_j is larger than its empirical error of more than ϵ:

$$\mathfrak{S}_j^\epsilon = \{S = \{(\mathbf{x}_1, y_1), \ldots, (\mathbf{x}_m, y_m)\} : \mathcal{E}(f_j) - E(f_j, S) > \epsilon\}$$

Given a $j \in \{1, \ldots, p\}$ and from the previous interpretation, the probability on the samples S of $\mathcal{E}(f_j) - E(f_j, S) > \epsilon$, is less than $e^{-2m\epsilon^2}$, that is

$$j \in \{1, \ldots, p\}; \mathbb{P}(\mathfrak{S}_j^\epsilon) \leq e^{-2m\epsilon^2} \qquad (2.10)$$

According to the equivalence (2.9), we also have:

$$\forall \epsilon > 0, \mathbb{P}\left(\max_{j \in \{1,\ldots,p\}} [\mathcal{E}(f_j) - E(f_j, S)] > \epsilon\right) = \mathbb{P}\left(\exists f \in \mathcal{F}, \mathcal{E}(f) - E(f, S) > \epsilon\right)$$

$$= \mathbb{P}\left(\mathfrak{S}_1^\epsilon \cup \ldots \cup \mathfrak{S}_p^\epsilon\right)$$

We now bound the previous equality by using the result of (2.10), and the union bound:

$$\forall \epsilon > 0, \mathbb{P}\left(\max_{j \in \{1,\ldots,p\}} [\mathcal{E}(f_j) - E(f_j, S)] > \epsilon\right)$$

$$= \mathbb{P}\left(\mathfrak{S}_1^\epsilon \cup \ldots \cup \mathfrak{S}_p^\epsilon\right) \leq \sum_{j=1}^{p} \mathbb{P}(\mathfrak{S}_j^\epsilon) \leq p e^{-2m\epsilon^2}$$

Solving this bound for $\delta = pe^{-2m\epsilon^2}$, i.e. $\epsilon = \sqrt{\frac{\ln(p/\delta)}{2m}}$, and by considering the opposite event we finally obtain:

$$\forall \delta \in]0,1], \mathbb{P}\left(\forall f \in \mathcal{F}, \mathcal{E}(f) \leq E(f,S) + \sqrt{\frac{\ln(|\mathcal{F}|/\delta)}{2m}}\right) \geq 1 - \delta \qquad (2.11)$$

Compared to the generalization bound obtained on a test set (Eq. 2.8), we see that the empirical error on a test set is a better estimator of the generalization bound; than the empirical error on a training set. In addition, the larger the size of a function class; the higher is the chance that the empirical error on the training set be a significant under-estimate of the generalization error.

Indeed, the interpretation of the bound (2.11), is that for a given $\delta \in]0,1]$ and for a fraction of possible training sets larger than $1 - \delta$; all the functions of the finite set $\mathcal{F}$ (including the function minimising the empirical error) have a generalization error less than their empirical error plus the residual term $\sqrt{\frac{\ln(|\mathcal{F}|/\delta)}{2m}}$. Furthermore, the difference in the worse case between the generalization and the empirical errors tends to 0 when the number of examples tends to infinity and this without making any particular assumption over the distribution of examples $\mathcal{D}$. Thus for any finite set of functions, the ERM principle is consistent for any probability distribution $\mathcal{D}$.

> **In Sum**
> The two steps that led to the generalization bound presented in the previous development are as follows:
>
> 1. For all fixed function $f_j \in \{f_1, \ldots, f_p\}$ and a given $\epsilon > 0$, bound the probability of $\mathcal{E}(f_j) - E(f_j, S) > \epsilon$, over the samples S;
> 2. Use the union bound to pass from the probability that holds for each single function in $\mathcal{F}$, to the probability that holds for all the functions in $\mathcal{F}$, at the same time.

2.2.2.2 Case of Infinite Sets of Functions

For the case of an infinite class of functions, the above approach is not directly applicable. Indeed; given a training set of m examples $S = \{(\mathbf{x}_1, y_1), \ldots, (\mathbf{x}_m, y_m)\}$, if we consider the following set:

$$\mathfrak{F}(\mathcal{F}, S) = \left\{ ((\mathbf{x}_1, f(\mathbf{x}_1)), \ldots, (\mathbf{x}_m, f(\mathbf{x}_m))) \mid f \in \mathcal{F} \right\} \qquad (2.12)$$

The size of this set corresponds to the number of possible ways that the functions of $\mathcal{F}$ can assign class labels to examples $(\mathbf{x}_1, \ldots, \mathbf{x}_m)$, and as these functions have two possible outputs (-1 or $+1$), the size of $\mathfrak{F}(\mathcal{F}, S)$ is finite, bounded by 2^m, and this for every considered class function $\mathcal{F}$. Thus a learning algorithm minimizing the empirical risk on a training set S, chooses the function among $|\mathfrak{F}(\mathcal{F}, S)|$ functions of $\mathcal{F}$ which performs class labels of S resulting in the smallest error. Thereby, there is only a finite number of functions that are involved into the estimation of the empirical error appearing in the expression of

$$\mathbb{P}\left(\sup_{f \in \mathcal{F}} [\mathcal{E}(f) - E(f, S)] > \epsilon\right) \qquad (2.13)$$

However, for a different training set S', the set $\mathfrak{F}(\mathcal{F}, S')$ will be different from $\mathfrak{F}(\mathcal{F}, S)$, and it is impossible to apply the bound for these sets by considering $|\mathfrak{F}(\mathcal{F}, S)|$.

The solution proposed by Vapnik and Chervonenkis is an elegant way to solve this problem, and it consists in replacing the generalization error $\mathcal{E}(f)$ in the expression (2.13) by the empirical error of f on another sample of the same size as S, and called virtual or ghost sample. Formally

Lemma 2.1 (Vapnik and Chervonenkis symmetrization lemma Vapnik (1999))
Let $\mathcal{F}$ be a function class (possibly infinite), and S and S' two training samples of the same size m. For every $\epsilon > 0$, such that $m\epsilon^2 \leq 2$ we have:

$$\mathbb{P}\left(\sup_{f \in \mathcal{F}} [\mathcal{E}(f) - E(f, S)] > \epsilon\right) \leq 2\mathbb{P}\left(\sup_{f \in \mathcal{F}} [E(f, S') - E(f, S)] > \epsilon/2\right) \quad (2.14)$$

Proof of the symmterization lemma
Let $\epsilon > 0$ and $f_S^* \in \mathfrak{F}(\mathcal{F}, S)$ the function that realises the sumpremum $\sup_{f \in \mathcal{F}} [\mathcal{E}(f) - E(f, S)]$. According to the remark above, f_S^* depends on the sample S. We have

$$\mathbb{1}_{[\mathcal{E}(f_S^*) - E(f_S^*, S) > \epsilon]} \mathbb{1}_{[\mathcal{E}(f_S^*) - E(f_S^*, S') < \epsilon/2]} = \mathbb{1}_{[\mathcal{E}(f_S^*) - E(f_S^*, S) > \epsilon \wedge E(f_S^*, S') - \mathcal{E}(f_S^*) \geq -\epsilon/2]}$$

$$\leq \mathbb{1}_{[E(f_S^*, S') - E(f_S^*, S) > \epsilon/2]}$$

By taking the expectation over S' on the previous inequality we have

$$\mathbb{1}_{[\mathcal{E}(f_S^*) - E(f_S^*, S) > \epsilon]} \mathbb{E}_{S' \sim \mathcal{D}^m} \mathbb{1}_{[\mathcal{E}(f_S^*) - E(f_S^*, S') < \epsilon/2]} \leq \mathbb{E}_{S' \sim \mathcal{D}^m} \mathbb{1}_{[E(f_S^*, S') - E(f_S^*, S) > \epsilon/2]}$$

That is

$$\mathbb{1}_{[\mathcal{E}(f_S^*) - E(f_S^*, S) > \epsilon]} \mathbb{P}(\mathcal{E}(f_S^*) - E(f_S^*, S') < \epsilon/2) \leq \mathbb{E}_{S' \sim \mathcal{D}^m} \mathbb{1}_{[E(f_S^*, S') - E(f_S^*, S) > \epsilon/2]}$$

For each example $(\mathbf{x}_i', y_i') \in S'$ denote by X_i the random variable $\frac{1}{m} e(f_S^*(\mathbf{x}_i'), y_i')$, as f_S^* is independent to the sample S', the random variables $X_i, i \in \{1, \ldots, m\}$ are all independent. The variance of the random variable

$E(f_S^*, S')$; $\mathbb{V}(E(f_S^*, S'))$; is thus equal to $\frac{1}{m}\mathbb{V}(e(f_S^*(\mathbf{x}'), y'))$ and according to Tchebychev (1867) inequality, we have

$$\mathbb{P}(\mathcal{E}(f_S^*) - E(f_S^*, S') \geq \epsilon/2) \leq \frac{4\mathbb{V}(e(f_S^*(\mathbf{x}'), y'))}{m\epsilon^2} \leq \frac{1}{m\epsilon^2}$$

The last inequality is due to the fact that $e(f_S^*(\mathbf{x}'), y'))$ is a random variable taking its values in $[0, 1]$ and that its variance is lower than $1/4$:

$$\left(1 - \frac{1}{m\epsilon^2}\right) \mathbb{1}_{[\mathcal{E}(f_S^*) - E(f_S^*, S) > \epsilon]} \leq \mathbb{E}_{S' \sim \mathcal{D}^m} \mathbb{1}_{[E(f_S^*, S') - E(f_S^*, S) > \epsilon/2]}$$

The result follows by taking the expectation over the sample S and by noting that $m\epsilon^2 \leq 2$, i.e. $\frac{1}{2} \leq (1 - \frac{1}{m\epsilon^2})$.

We note that the expectation appearing in the left inequality of (2.14) is with respect to the distribution of an i. i. d. sample of size m, while the one in the right is following the distribution of an i. i. d. sample of size $2m$. The extension of the generalization error for an infinite function class, $\mathcal{F}$, is done by studying the largest difference between the empirical losses of functions in $\mathcal{F}$ on any two training sets of the same size. In fact, the important quantity that occurs in the previous result is the maximum number of possible labelings assigned by functions in $\mathcal{F}$ to a sample of size $2m$, denoted by $\mathfrak{G}(\mathcal{F}, 2m)$, where

$$\mathfrak{G}(\mathcal{F}, m) = \max_{S \in \mathcal{X}^m} |\mathfrak{F}(\mathcal{F}, S)| \tag{2.15}$$

$\mathfrak{G}(\mathcal{F}, m)$ is called the growth function and it can be seen as a measure of the size of a function class $\mathcal{F}$, as it is shown in the following result:

Theorem 2.1 (Vapnik and Cervonenkis (1971); Vapnik (1999), Chap. 3) *Let $\delta \in]0, 1]$ and S be a training set of size m generated i. i. d. with respect to a probability distribution $\mathcal{D}$; we have with a probability of at least $1 - \delta$:*

$$\forall f \in \mathcal{F}, \mathcal{E}(f) \leq E(f, S) + \sqrt{\frac{8 \ln (\mathfrak{G}(\mathcal{F}, 2m)) + 8 \ln (4/\delta)}{m}} \tag{2.16}$$

Proof of theorem 2.1

Let ϵ be a positive real value, from the symmetrisation lemma 2.1, we have

$$\mathbb{P}\left(\sup_{f \in \mathcal{F}} [\mathcal{E}(f) - E(f, S)] > \epsilon\right) \leq 2\mathbb{P}\left(\sup_{f \in \mathcal{F}} [E(f, S') - E(f, S)] > \epsilon/2\right)$$

$$= 2\mathbb{P}\left(\sup_{f \in \mathfrak{F}(\mathcal{F}, S \cup S')} [E(f, S') - E(f, S)] > \epsilon/2\right)$$

Which according to the union bound gives

$$\mathbb{P}\left(\sup_{f \in \mathcal{F}}[\mathcal{E}(f) - E(f, S)] > \epsilon\right)$$

$$\leq 2\mathfrak{G}(\mathcal{F}, 2m)\mathbb{P}\left([E(f, S') - E(f, S)] > \epsilon/2\right)$$

From Hoeffding (1963) inequality we have

$$\mathbb{P}\left([E(f, S') - E(f, S)] > \epsilon\right) \leq 2e^{-m\epsilon^2/2}$$

That is

$$\mathbb{P}\left(\sup_{f \in \mathcal{F}}[\mathcal{E}(f) - E(f, S)] > \epsilon\right) \leq 4\mathfrak{G}(\mathcal{F}, 2m)e^{-m\epsilon^2/8}$$

The result follows by resolving $4\mathfrak{G}(\mathcal{F}, 2m)e^{-m\epsilon^2/8} = \delta$ with respect to ϵ.

The important result of this theorem is that the ERM principle is consistent in the case where $\sqrt{\frac{\ln(\mathfrak{G}(\mathcal{F},2m))}{m}}$ tends to 0 when m tends to infinity. Furthermore, as the distribution of observations $\mathcal{D}$ is not involed in the definition of the growth function, the previous analysis is valid for any $\mathcal{D}$. Thus, a sufficient condition of the consistency of the ERM principal, for all probability distributions $\mathcal{D}$ and any infinite function class, is $\lim_{m \to \infty} \sqrt{\frac{\ln(\mathfrak{G}(\mathcal{F}, 2m))}{m}} = 0$. On the other hand, $\mathfrak{G}(\mathcal{F}, m)$ cannot be estimated and the only certainty that we have is that it is bounded by 2^m. Furthermore, in the case where the growth function attains this bound, $\mathfrak{G}(\mathcal{F}, m) = 2^m$, we know that there exists a sample of size m such that the function class $\mathcal{F}$ is able to generate all the possible labelings of this sample, we then say that the sample is shattered by $\mathcal{F}$. From this observation, Vapnik and Chervonenkis proposed an auxiliary quantity, called VC dimension, to study the growth function and which is defined as following.

Definition 2.1 (VC Dimension, Vapnik (1999)) Let $\mathcal{F} = \{f : \mathcal{X} \to \{-1, +1\}\}$ be a discrete valued function class. The VC dimension of $\mathcal{F}$ is the largest integer $\mathcal{V}$ verifying $\mathfrak{G}(\mathcal{F}, \mathcal{V}) = 2^{\mathcal{V}}$. In other terms, $\mathcal{V}$ is the largest number of points that the class of functions can shatter. If such integer does not exist, VC dimension of $\mathcal{F}$ is considered as infinite.

Figure 2.2 illustrates the estimation of the VC dimension of a class of linear functions in the plan. From the previous definition, we see that the higher the VC dimension, $\mathcal{V}$, of a function class, the larger the growth function $\mathfrak{G}(\mathcal{F}, m)$ of this class, and that for any $m \geq \mathcal{V}$. An important property of the VC dimension known as

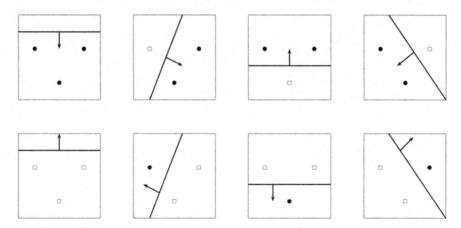

Fig. 2.2 Shatering of points in a plan of dimension $d = 2$ by a class of linear functions. Each classifier separates the plan in two sub planes, with a normal vector pointing to the subspace containing positive examples (represented by *circles*). The maximum number of points in the plan that can be shattered by this class, or the VC dimension of this class, is in this case equal to 3

the Sauer (1972) lemma, is that the VC dimension of a function class is a measure of its capacity and it is formally stated as following.

Lemma 2.2 (Vapnik and Cervonenkis (1971); Sauer (1972); Shelah (1972)[2])
Let $\mathcal{F}$ be a class of functions with values in $\{-1, +1\}$ with a finite VC dimension, $\mathcal{V}$. For all m, the growth function $\mathfrak{G}(\mathcal{F}, m)$ is bounded by

$$\mathfrak{G}(\mathcal{F}, m) \leq \sum_{i=0}^{\mathcal{V}} \binom{m}{i} \tag{2.17}$$

And for all $m \geq \mathcal{V}$

$$\mathfrak{G}(\mathcal{F}, m) \leq \left(\frac{m}{\mathcal{V}}\right)^{\mathcal{V}} e^{\mathcal{V}} \tag{2.18}$$

There exist different profs of this lemma (Sauer 1972; Shelah 1972; Bronnimann and Goodrich 1995; Cesa-Bianchi and Haussler 1998; Mohri et al. 2012), including the one based on induction over $m + \mathcal{V}$ that we present next. First note that the inequality (2.17) is true for $\mathcal{V} = 0$ and $m = 0$, in fact:

- If $\mathcal{V} = 0$, it means that the function class is not able to shatter any points, i.e. it always provides the same labeling that is $\mathfrak{G}(\mathcal{F}, m) = 1 = \binom{m}{0}$.
- if $m = 0$, we are in the presence of a trivial labeling of the empty set, i.e. $\mathfrak{G}(\mathcal{F}, 0) = 1 = \sum_{i=0}^{\mathcal{V}} \binom{0}{i}$

---

[2] The lemma was first stated, and in a slightly different form, in Vapnik and Cervonenkis (1971).

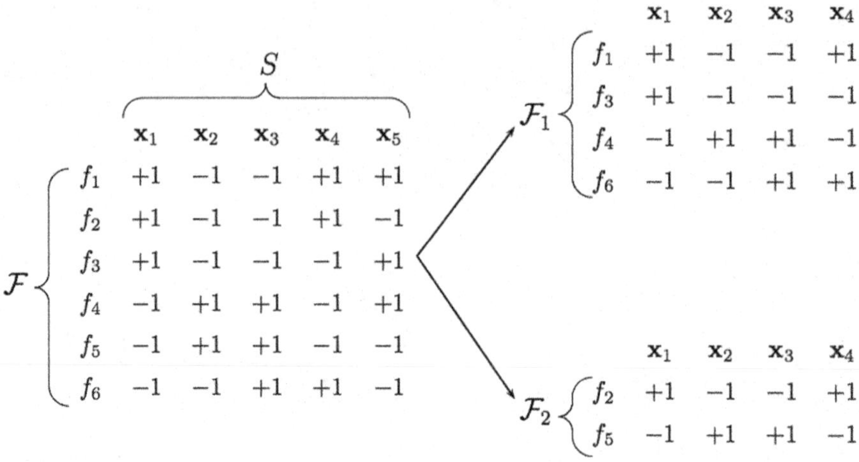

Fig. 2.3 Construction of the sets $\mathcal{F}_1$ and $\mathcal{F}_2$ from the class of functions $\mathcal{F}$ for the proof of Sauer (1972) lemma on a toy example

Suppose now that the inequality (2.17) is true for $m' + \mathcal{V}' < m + \mathcal{V}$, and given a training set $S = \{\mathbf{x}_1, \ldots, \mathbf{x}_m\}$ and a class of functions $\mathcal{F}$ with VC dimension , $\mathcal{V}$; let

$$\text{prove } |\mathfrak{F}(\mathcal{F}, S)| \leq \sum_{i=0}^{\mathcal{V}} \binom{m}{i}.$$

Consider two subsets of functions $\mathcal{F}_1$ and $\mathcal{F}_2$, from $\mathcal{F}$, defined on the training set $S' = S \backslash \{\mathbf{x}_m\}$ of size $m-1$. Let build the function class $\mathcal{F}_1$, by adding all the functions in $\mathcal{F}$ such that the prediction vectors of these foncions on S' are all different, and let $\mathcal{F}_2 = \mathcal{F} \backslash \mathcal{F}_1$. Thus, if two functions of $\mathcal{F}$ have the same prediction vector on S' and that the only difference between them lies in their predictions on the example $\mathbf{x}_m$, one of these functions will be put in $\mathcal{F}_1$ and the other in $\mathcal{F}_2$. Figure 2.3 illustrates this construction for a toy problem. The pairs of functions (f_1, f_2) and (f_4, f_5) have the same prediction vectors on $S' = S \backslash \{\mathbf{x}_5\}$, and the sets $\mathcal{F}_1, \mathcal{F}_2$ will contain each one of the functions of these pairs.

We note that if one set is shattered by the class $\mathcal{F}_1$, it will also be shattered by the class $\mathcal{F}$ since $\mathcal{F}_1$ contains all non-redundant functions of $\mathcal{F}$ on S', thus

$$\text{VC dimension}(\mathcal{F}_1) \leq \text{VC dimension}(\mathcal{F}) = \mathcal{V}$$

Further, if a set S' is shattered by $\mathcal{F}_2$, the set $S' \cup \{\mathbf{x}_m\}$ will be shattered by $\mathcal{F}$ since, for all functions in $\mathcal{F}_2$, $\mathcal{F}$ also contains the other function having a different prediction than the first function on $\mathbf{x}_m$, thus VC dimension$(\mathcal{F}) \geq$ VC dimension $(\mathcal{F}_2) + 1$ thereby

$$\text{VC dimension}(\mathcal{F}_2) \leq \mathcal{V} - 1$$

According to the hypothesis of the induction we have:

$$|\mathcal{F}_1| = |\mathfrak{F}(\mathcal{F}_1, S')| \le \mathfrak{G}(\mathcal{F}_1, m-1) \le \sum_{i=0}^{v} \binom{m-1}{i}$$

$$|\mathcal{F}_2| = |\mathfrak{F}(\mathcal{F}_2, S')| \le \mathfrak{G}(\mathcal{F}_2, m-1) \le \sum_{i=0}^{v-1} \binom{m-1}{i}$$

The demonstration ends after a variable change and the recursive, purely additive, formula

$$|\mathfrak{F}(\mathcal{F}, S)| = |\mathcal{F}_1| + |\mathcal{F}_2|$$

$$\le \sum_{i=0}^{v} \binom{m-1}{i} + \sum_{i=0}^{v-1} \binom{m-1}{i}$$

$$= \sum_{i=0}^{v} \binom{m-1}{i} + \sum_{i=0}^{v} \binom{m-1}{i-1}$$

$$= \sum_{i=0}^{v} \binom{m}{i}$$

Thus, as the previous inequality holds for any set S of size m, we have $\mathfrak{G}(\mathcal{F}, m) \le \sum_{i=0}^{v} \binom{m}{i}$.

To proof the inequality (2.18) we make use of the binomial formula. Thus, from the inequality (2.17) and in the case where $\frac{v}{m} \le 1$ we have

$$\left(\frac{v}{m}\right)^{v} \mathfrak{G}(\mathcal{F}, m) \le \left(\frac{v}{m}\right)^{v} \sum_{i=0}^{v} \binom{m}{i} \le \sum_{i=0}^{v} \left(\frac{v}{m}\right)^{i} \binom{m}{i}$$

By multiplying the term in the right by $1^{m-i} = 1$ and using the binomial formula, it comes

$$\left(\frac{v}{m}\right)^{v} \mathfrak{G}(\mathcal{F}, m) \le \sum_{i=0}^{v} \binom{m}{i} \left(\frac{v}{m}\right)^{i} 1^{m-i}$$

$$= \left(1 + \frac{v}{m}\right)^{m}$$

Finally from the inequality $\forall z \in \mathbb{R}, (1-z) \le e^{-z}$ we have:

$$\mathfrak{G}(\mathcal{F}, m) \le \left(\frac{m}{v}\right)^{v} \left(1 + \frac{v}{m}\right)^{m} \le \left(\frac{m}{v}\right)^{v} e^{v}$$

From the previous result, we can see that the values taken by the growth function associated to the function class $\mathcal{F}$, will depend on the existence or not of the VC dimension of $\mathcal{F}$:

$$\forall m, \mathfrak{G}(\mathcal{F}, m) = \begin{cases} O(m^{\mathcal{V}}), & \text{if } \mathcal{V} \text{ is finite,} \\ 2^m, & \text{if } \mathcal{V} \text{ is infinite.} \end{cases}$$

Furthermore, in the case where the VC dimension, $\mathcal{V}$, of the function class $\mathcal{F}$ is finite and that we have enough training examples such that $m \geq \mathcal{V}$, the evolution of the growth function becomes polynomial with respect to m, i.e. $\ln \mathfrak{G}(\mathcal{F}, 2m) \leq \mathcal{V} \ln \frac{2em}{\mathcal{V}}$ (2.18). This result allows to have a new expression for the generalization bound of (2.16) and which is calculable for a known $\mathcal{V}$ and any given training set.

Corollary 2.1 (*Generalization bound using the VC dimension*). *Let $\mathcal{X} \in \mathbb{R}^d$ be a vectoriel space, $\mathcal{Y} = \{-1, +1\}$ an output space and $\mathcal{F}$ a class of functions with values in $\mathcal{Y}$ and VC dimension, $\mathcal{V}$. Suppose that the pairs of examples $(\mathbf{x}, y) \in \mathcal{X} \times \mathcal{Y}$ are generated i. i. d. with respect to a probability distribution $\mathcal{D}$. For any $\delta \in]0, 1]$, for all $f \in \mathcal{F}$ and all training set $S \in (\mathcal{X} \times \mathcal{Y})^m$ of size $m \geq \mathcal{V}$ also generated i. i. d. with respect to the same probability distribution, the following inequality holds with a probability at least $1 - \delta$:*

$$\mathcal{E}(f) \leq E(f, S) + \sqrt{\frac{8\mathcal{V} \ln \frac{2em}{\mathcal{V}} + 8 \ln \frac{4}{\delta}}{m}} \tag{2.19}$$

Thus, when $\lim\limits_{m \to \infty} \dfrac{8\mathcal{V} \ln \frac{2em}{\mathcal{V}} + 8 \ln \frac{4}{\delta}}{m} = 0$, we can deduce from the previous result a sufficient condition for the consistency of the ERM principle and which can be stated as *for a given binary function class $\mathcal{F}$, if the VC dimension of $\mathcal{F}$ is finite, then, the ERM principle is consistent for all generating probability distributions $\mathcal{D}$.*

For the consistency of the ERM principle, Vapnik (1999) showed that it is also necessary that the VC dimension of the considered function be finite, and that for any probability distribution $\mathcal{D}$. Thus we have the principal result in machine learning which states

CENTRAL NOTION. Necessary and sufficient condition for the consistency of the ERM principle

For any probability distribution generating the examples, the ERM principal is consistent if and only if the VC dimension of the considered function class is finite.

2.2.3 Structural Risk Minimization

From the previous study, we see that the larger (respectively the smaller) the capacity of a function class, the higher (respectively the lower) the possibility of assigning different labels to examples in a training set, and the lower (respectively the higher) the empirical error of a function from this class on the considered training set, and this without being able to guarantee a low generalization error. The difficulty of learning is hence to minimize the bound on a generalization error by realizing a compromise

---

Input :
- A prediction problem issue from an application domain.

Procedure:

- With a prior knowledge on the application domain, choose a function class (for example polynomial functions);
- Divide the class of functions in a nested hierarchy of subclasses (for example poylnoms of increasing degree);
- On a given training set, learn a prediction function from each of the subclasses according to the ERM principle ;

Output : From all the learned functions, select the one for which we obtain the best bound of its generalization error (function realising the best compromise).

Algorithm 1: Structural risk minimization principle (Vapnik and Cervonenkis, 1974)

---

between a low empirical error and a low capacity of the function class. Figure 2.4 illustrates this compromise which is commonly called structural risk minimization (Vapnik and Cervonenkis 1974) and which states as following (algorithme 1).

In Sum
1. For induction, we should control the capacity of the class of functions.
2. The ERM principle is consistent for any probability distribution generating the examples, if and only if, the VC dimension of the class of functions that is considered is finite.
3. The study of the consistency of the ERM principle led to the second fundamental principle of machine learning called structural risk minimization (SRM).
4. Learning is a compromise between a low empirical risk and a high capacity of the class of functions in use.

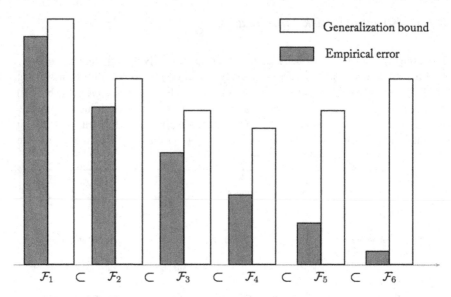

Generalization bound

Empirical error

$\mathcal{F}_1 \subset \mathcal{F}_2 \subset \mathcal{F}_3 \subset \mathcal{F}_4 \subset \mathcal{F}_5 \subset \mathcal{F}_6$

Fig. 2.4 Illustration of the structural risk minimization principle. The abscissa shows a hierarchy of nested subclass of functions with an increasing capacity, from *left* to *right*. The larger the capacity of a class of functions, the lower the empirical error of a function from this class in a training set and the worst its generalization error. The structural risk minimization consists in selecting the function from the function class for which we have the best estimate of its generalization error

2.3   Data-Dependent Generalization Error Bounds

We have seen that the growth function and the VC dimension are two measures for quantifying the capacity of a class of functions independently to the unknown distribution probability generating the data. However, the growth function is practically impossible to estimate, and in the majority of cases the VC dimension of a function class is too large to allow having an accurate estimate of the generalization error. There are other quantities leading to more accurate estimates of the capacity of a class of functions and which are dependent to the training sample. Among the existing quantities, we will focus in this section on the Rademacher complexity introduced by Koltchinskii and Panchenko (2000); Koltchinskii (2001) and which has opened an active research field in learning theory.

2.3.1   Rademacher Complexity

The empirical Rademacher complexity estimates the richness of a function class $\mathcal{F}$ by measuring the degree to which the latter is able fit to random noise on a training set $S = \{(\mathbf{x}_1, y_1), \ldots, (\mathbf{x}_m, y_m)\}$ of size m generated i. i. d. with respect to a probability distribution $\mathcal{D}$. This complexity is estimated through Rademacher variables $\boldsymbol{\sigma} = (\sigma_1, \ldots, \sigma_m)^\top$ which are independent discrete random variables taking values in $\{-1, +1\}$ with the same probability $1/2$, i.e. $\forall i \in \{1, \ldots, m\}$; $\mathbb{P}(\sigma_i = -1) = \mathbb{P}(\sigma_i = +1) = 1/2$, and is defined as:

$$R(\mathcal{F}, S) = \frac{2}{m} \mathbb{E}_\sigma \left[\sup_{f \in \mathcal{F}} \left| \sum_{i=1}^m \sigma_i f(\mathbf{x}_i) \right| \mid \mathbf{x}_1, \ldots, \mathbf{x}_m \right] \tag{2.20}$$

Where $\mathbb{E}_\sigma$ is the conditionnel expectation with respect to the data, and taken over the Rademacher variables. In this expression, $\left| \sum_{i=1}^m \sigma_i f(\mathbf{x}_i) \right|$ is the absolute value of the dot product between a random noise vector $\boldsymbol{\sigma}$ and the vector of class predictions induced by f, and which estimates the correlation between the vectors of predictions and the random noise. Thus, the supremum $\sup_{f \in \mathcal{F}} \left| \sum_{i=1}^m \sigma_i f(\mathbf{x}_i) \right|$ estimes to which degree the class of functions $\mathcal{F}$ is correlated to the noise vector $\boldsymbol{\sigma}$ over the sample S, and the empirical Rademacher complexity is an average of this estimation. Thus, this complexity describes the richness of the function class $\mathcal{F}$; the larger the capacity of this class and the higher the chance of finding a function f_S which would be correlated to the noise on S. Furthermore, we define the Rademacher complexity of the class of functions $\mathcal{F}$ independently to a given training set by taking the expectation of the empirical Rademacher complexity of the class over all samples of size m generated i. i. d. with respect to $\mathcal{D}$:

$$\mathcal{R}_m(\mathcal{F}) = \mathbb{E}_{S \sim \mathcal{D}^m} R(\mathcal{F}, S) = \frac{2}{m} \mathbb{E}_{S\sigma} \left[\sup_{f \in \mathcal{F}} \left| \sum_{i=1}^m \sigma_i f(\mathbf{x}_i) \right| \right] \tag{2.21}$$

2.3.2   Link Between the Rademacher Complexity and the VC Dimension

There exists a link between the Rademacher complexity, $\mathcal{R}_m(\mathcal{F})$, of a class of functions $\mathcal{F}$ and its associated growth function $\mathfrak{G}(\mathcal{F}, m)$, and hence to its VC dimension in the case where the inequality (2.18) holds. This link follows the result presented in (Massart 2000, lemma 5.2), and is known as Massart's lemma:

Lemma 2.3 (Massart (2000))

Let $\mathcal{A}$ be a finite subset of $\mathbb{R}^m$, and $\boldsymbol{\sigma} = (\sigma_1, \ldots, \sigma_m)^\top$, m independent Rademacher random variables, the following inequality always holds:

$$\mathbb{E}_\sigma \left[\sup_{a \in \mathcal{A}} \sum_{i=1}^m \sigma_i a_i \right] \leq r \sqrt{2 \ln |\mathcal{A}|} \tag{2.22}$$

where $r = \sup_{a \in \mathcal{A}} \|a\|$.

Proof of Massart (2000) lemma
From Jensen's inequality, we have for all real value $\lambda > 0$

$$\exp\left(\lambda \mathbb{E}_\sigma\left[\sup_{a \in \mathcal{A}} \sum_{i=1}^m \sigma_i a_i\right]\right) \leq \mathbb{E}_\sigma\left[\exp\left(\lambda \sup_{a \in \mathcal{A}} \sum_{i=1}^m \sigma_i a_i\right)\right]$$

$$= \mathbb{E}_\sigma\left[\sup_{a \in \mathcal{A}} \exp\left(\lambda \sum_{i=1}^m \sigma_i a_i\right)\right]$$

$$\leq \mathbb{E}_\sigma\left[\sum_{a \in \mathcal{A}} \exp\left(\lambda \sum_{i=1}^m \sigma_i a_i\right)\right]$$

$$= \sum_{a \in \mathcal{A}} \mathbb{E}_\sigma\left[\exp\left(\lambda \sum_{i=1}^m \sigma_i a_i\right)\right]$$

$$= \sum_{a \in \mathcal{A}} \mathbb{E}_\sigma \prod_{i=1}^m \left[\exp(\lambda \sigma_i a_i)\right]$$

as the random variables $\sigma_1, \ldots, \sigma_m$ are independent, and by definition $\forall i, \mathbb{P}(\sigma_i = -1) = \mathbb{P}(\sigma_i = +1) = 1/2$, we have

$$\exp\left(\lambda \mathbb{E}_\sigma\left[\sup_{a \in \mathcal{A}} \sum_{i=1}^m \sigma_i a_i\right]\right) \leq \sum_{a \in \mathcal{A}} \prod_{i=1}^m \mathbb{E}_{\sigma_i}\left[\exp(\lambda \sigma_i a_i)\right]$$

$$= \sum_{a \in \mathcal{A}} \prod_{i=1}^m \frac{e^{-\lambda a_i} + e^{\lambda a_i}}{2}$$

Using the inequality $\forall z \in \mathbb{R}, \frac{e^{-z} + e^z}{2} \leq e^{z^2/2}$, we get

$$\exp\left(\lambda \mathbb{E}_\sigma\left[\sup_{a \in \mathcal{A}} \sum_{i=1}^m \sigma_i a_i\right]\right) \leq \sum_{a \in \mathcal{A}} \prod_{i=1}^m e^{\lambda^2 a_i^2/2}$$

$$= \sum_{a \in \mathcal{A}} e^{\lambda^2 \|a\|^2/2} \leq |\mathcal{A}| e^{\lambda^2 r^2/2}$$

Thus, by taking the logarithm and dividing by λ it comes:

$$\mathbb{E}_\sigma\left[\sup_{a \in \mathcal{A}} \sum_{i=1}^m \sigma_i a_i\right] \leq \frac{\ln |\mathcal{A}|}{\lambda} + \frac{\lambda r^2}{2} \tag{2.23}$$

As this inequality holds for all $\lambda > 0$, it also holds for $\lambda^* = \frac{\sqrt{2\ln|\mathcal{A}|}}{r}$ which minimizes the term in the right hand side of the inequality. The result (2.22) follows by plugging the value of λ^* into (2.23). Further, by considering the inequality (2.22), we also have:

$$\mathbb{E}_\sigma\left[\sup_{a\in\mathcal{A}}\left|\sum_{i=1}^m\sigma_i a_i\right|\right] = \mathbb{E}_\sigma\left[\sup_{a\in\mathcal{A}\cup-\mathcal{A}}\sum_{i=1}^m\sigma_i a_i\right] \le r\sqrt{2\ln(2|\mathcal{A}|)} \qquad (2.24)$$

With this result we can bound the Rademacher complexity of a class of functions with respect to the associated growth function.

Corollary 2.2 *Let $\mathcal{F}$ be a class of functions with values in $\{-1,+1\}$, and a finite VC dimension, $\mathcal{V}$, and $\mathfrak{G}(\mathcal{F},m)$the associated growth function. For all integer greater than zero m, the Rademacher complexity $\mathcal{R}_m(\mathcal{F})$ is bounded by*

$$\mathcal{R}_m(\mathcal{F}) \le \sqrt{\frac{8\ln(2\mathfrak{G}(\mathcal{F},m))}{m}} \qquad (2.25)$$

And, in the case where $m \ge \mathcal{V}$, we have

$$\mathcal{R}_m(\mathcal{F}) \le \sqrt{8\left(\frac{\ln 2}{m} + \frac{\mathcal{V}}{m}\ln\frac{em}{\mathcal{V}}\right)} \qquad (2.26)$$

Proof of corollary 2.2
For a given training set $S = \{(\mathbf{x}_1, y_1), \ldots, (\mathbf{x}_m, y_m)\}$ with size m, as the functions $f \in \mathcal{F}$ have values in $\{-1,+1\}$, the norm of the vectors $(f(\mathbf{x}_1),\ldots,f(\mathbf{x}_m))^\top$ for all $f \in \mathcal{F}$ is bounded by $\sqrt{m}$. From the inequality (2.24), the definition (2.12) and the definition of the growth function (2.15), we have

$$\mathcal{R}_m(\mathcal{F}) = \frac{2}{m}\mathbb{E}_S\left[\mathbb{E}_\sigma\left[\sup_{f\in\mathfrak{F}(\mathcal{F},S)}\left|\sum_{i=1}^m\sigma_i f(\mathbf{x}_i)\right|\right]\right]$$
$$\le \frac{2}{m}\mathbb{E}_S\left[\sqrt{2m\ln(2\mathfrak{G}(\mathcal{F},m))}\right] \le \sqrt{\frac{8\ln(2\mathfrak{G}(\mathcal{F},m))}{m}}$$

The inequality (2.26) follows from the previous result and the second part of Sauer's lemma (2.18).

2.3.3 Different Steps for Obtaining a Generalization Bound with the Rademacher Complexity

In this section, we give an outline on how to obtain a generalization bound in the case of binary classification with i. i. d. examples and using the theory developed around the Rademacher complexity (Bartlett and Mendelson (2003a); Taylor and Cristianini (2004)—Chap. 4):

Theorem 2.2 *Let $\mathcal{X} \in \mathbb{R}^d$ be a vectoriel space and $\mathcal{Y} = \{-1, +1\}$ an output space. Suppose that the pairs of examples $(\mathbf{x}, y) \in \mathcal{X} \times \mathcal{Y}$ are generated i. i. d. with respect to the distribution probability $\mathcal{D}$. Let $\mathcal{F}$ be a class of functions having values in $\mathcal{Y}$ and $e : \mathcal{Y} \times \mathcal{Y} \to [0, 1]$ a given instantaneous loss. For all $\delta \in]0, 1]$, we have for all $f \in \mathcal{F}$ and all training set of size m, $S \in (\mathcal{X} \times \mathcal{Y})^m$ also generated i. i. d. with respect to the same probability distribution, then with probability at least $1 - \delta$ we have:*

$$\mathcal{E}(f) \leq E(f, S) + \mathcal{R}_m(L \circ \mathcal{F}) + \sqrt{\frac{\ln \frac{1}{\delta}}{2m}} \qquad (2.27)$$

and also with probability at least $1 - \delta$

$$\mathcal{E}(f) \leq E(f, S) + R(L \circ \mathcal{F}, S) + 3\sqrt{\frac{\ln \frac{2}{\delta}}{2m}} \qquad (2.28)$$

Where $L \circ \mathcal{F} = \{(\mathbf{x}, y) \mapsto L(f(\mathbf{x}), y) \mid f \in \mathcal{F}\}$.

The main interest of this theorem appears in the second bound which involves the empirical Rademacher complexity of the class of functions $L \circ \mathcal{F}$. We will see later that for certain class of functions it is indeed simple the associated empirical Rademacher complexity, and thus have a generalization bound estimable on any given training set.

We will present in the following, the three major steps that achieve these bounds.

1. Link the supremum of $\mathcal{E}(f) - E(f, S)$ on $\mathcal{F}$ with its expectation

As previously, we are looking for a uniform bound which holds for all functions f of a given class $\mathcal{F}$ as well as all training sets S, noting that

$$\forall f \in \mathcal{F}, \forall S, \mathcal{E}(f) - E(f, S) \leq \sup_{f \in \mathcal{F}} [\mathcal{E}(f) - E(f, S)] \qquad (2.29)$$

The study of this bound is achieved by linking the supremum appearing, in the right hand side of the above inequality, with its expectation through a powerful tool developed for empirical processes (Vapnik and Cervonenkis 1971; Blumer et al. 1989; Ehrenfeucht et al. 1989; Giné 1996) by McDiarmid (1989), and known as the theorem of bounded differences, or McDiarmid (1989) inequality .

Theorem 2.3 (McDiarmid (1989) inequality). *Let* $I \subset \mathbb{R}$ *be a real valued interval, and* $(X_1, ..., X_m)$, *m independent random variables taking values in* I^m. *Let* Φ : $I^m \to \mathbb{R}$ *be defined such that:* $\forall i \in \{1, ..., m\}, \exists c_i \in \mathbb{R}$ *the following inequality holds for any* $(x_1, ..., x_m) \in I^m$ *and* $\forall x' \in I$:

$$|\Phi(x_1, .., x_{i-1}, x_i, x_{i+1}, .., x_m) - \Phi(x_1, .., x_{i-1}, x', x_{i+1}, .., x_m)| \leq c_i$$

We have then

$$\forall \epsilon > 0, \mathbb{P}(\Phi(x_1, ..., x_m) - \mathbb{E}[\Phi] > \epsilon) \leq e^{\frac{-2\epsilon^2}{\sum_{i=1}^m c_i^2}}$$

Thus, by considering the following function

$$\Phi : S \mapsto \sup_{f \in \mathcal{F}} [\mathcal{E}(f) - E(f, S)]$$

It may be noted that for two samples S and S^i of size m containing exactly the same examples except the pair $(x_i, y_i) \in S$ instead of which S^i contains the pair (x', y') sampled according to the same probability distribution $\mathcal{D}$, the difference $|\Phi(S) - \Phi(S^i)|$ is then bounded by $c_i = 1/m$ as the loss function e (involved in $\mathcal{E}$ and E) has values in $[0, 1]$. McDiarmid (1989) inequality can then be applied for the function Φ with $c_i = 1/m, \forall i$, thus:

$$\forall \epsilon > 0, \mathbb{P}\left(\sup_{f \in \mathcal{F}} [\mathcal{E}(f) - E(f, S)] - \mathbb{E}_S \sup_{f \in \mathcal{F}} [\mathcal{E}(f) - E(f, S)] > \epsilon\right) \leq e^{-2m\epsilon^2}$$

2. Bound $\mathbb{E}_S \sup_{f \in \mathcal{F}} [\mathcal{E}(f) - E(f, S)]$ **with respect to** $\mathcal{R}_m(L \circ \mathcal{F})$

This step is a symmetrisation step and it consists in introducing a second virtual sample S' also generated i. i. d. with respect to $\mathcal{D}^m$ into $\mathbb{E}_S \sup_{f \in \mathcal{F}} [\mathcal{E}(f) - E(f, S)]$. This second sample plays a symmetric role with respect to S, and its introduction into the expectation of the supremum constitutes the most technical development with respect to those used in the three aforementioned steps.

Proof Bounding $\mathbb{E}_S \sup_{f \in \mathcal{F}} [\mathcal{E}(f) - E(f, S)]$

The bounding of $\mathbb{E}_S \sup_{f \in \mathcal{F}} [\mathcal{E}(f) - E(f, S)]$ begins by noting that for any function $f \in \mathcal{F}$, the expectation of f on any virtual sample S' is an unbiased estimator of its generalization error $\mathcal{E}(f)$, i.e. $\mathcal{E}(f) = \mathbb{E}_{S'} E(f, S')$ and that $E(f, S) = \mathbb{E}_{S'} E(f, S)$. Thus, we have:

$$\mathbb{E}_S \sup_{f \in \mathcal{F}} (\mathcal{E}(f) - E(f, S)) = \mathbb{E}_S \sup_{f \in \mathcal{F}} [\mathbb{E}_{S'} (E(f, S') - E(f, S))]$$

$$\leq \mathbb{E}_S \mathbb{E}_{S'} \sup_{f \in \mathcal{F}} [\mathcal{E}(f, S') - E(f, S)]$$

The previous inequality is due to the fact that the supremum of the expectation is lower than the expectation of the supremum. The second point in this step consists in introducing the Rademacher variables into the supremum:

$$\mathbb{E}_S \mathbb{E}_{S'} \sup_{f \in \mathcal{F}} \left[\frac{1}{m} \sum_{i=1}^{m} \sigma_i (L(f(\mathbf{x}_i'), y_i') - L(f(\mathbf{x}_i), y_i)) \right] \qquad (2.30)$$

For a fixed i, $\sigma_i = 1$ does not change anything but $\sigma_i = -1$ consists in swapping both examples $(\mathbf{x}_i', y_i')$ and $(\mathbf{x}_i, y_i)$. Thus, when we take the expectations over S and S', the introduction of Rademacher variables does not change anything. This is true for all $i \in \{1, \ldots, m\}$, thus by taking the expectation over $\boldsymbol{\sigma} = (\sigma_1, \ldots, \sigma_m)$, we get:

$$\mathbb{E}_S \mathbb{E}_{S'} \sup_{f \in \mathcal{F}} [\mathcal{E}(f, S') - E(f, S)]$$

$$= \mathbb{E}_S \mathbb{E}_{S'} \mathbb{E}_{\sigma} \sup_{f \in \mathcal{F}} \left[\frac{1}{m} \sum_{i=1}^{m} \sigma_i (L(f(\mathbf{x}_i'), y_i') - L(f(\mathbf{x}_i), y_i)) \right]$$

By applying the triangular inequality $\sup = \|.\|_\infty$ it comes

$$\mathbb{E}_S \mathbb{E}_{S'} \mathbb{E}_{\sigma} \sup_{f \in \mathcal{F}} \left[\frac{1}{m} \sum_{i=1}^{m} \sigma_i (L(f(\mathbf{x}_i'), y_i') - L(f(\mathbf{x}_i), y_i)) \right]$$

$$\leq \mathbb{E}_S \mathbb{E}_{S'} \mathbb{E}_{\sigma} \sup_{f \in \mathcal{F}} \frac{1}{m} \sum_{i=1}^{m} \sigma_i L(f(\mathbf{x}_i'), y_i')$$

$$+ \mathbb{E}_S \mathbb{E}_{S'} \mathbb{E}_{\sigma} \sup_{f \in \mathcal{F}} \frac{1}{m} \sum_{i=1}^{m} (-\sigma_i) L(f(\mathbf{x}_i'), y_i')$$

Finally as $\forall i$, σ_i and $-\sigma_i$ have the same distribution we have

$$\mathbb{E}_S \mathbb{E}_{S'} \sup_{f \in \mathcal{F}} [\mathcal{E}(f, S') - E(f, S)] \leq 2 \underbrace{\mathbb{E}_S \mathbb{E}_{\sigma} \sup_{f \in \mathcal{F}} \frac{1}{m} \sum_{i=1}^{m} \sigma_i L(f(\mathbf{x}_i), y_i)}_{\leq \mathcal{R}_m(L \circ \mathcal{F})} \quad (2.31)$$

In summarizing the results obtained so far, we have:

1. $\forall f \in \mathcal{F}, \forall S, \mathcal{E}(f) - E(f, S) \le \sup_{f \in \mathcal{F}} [\mathcal{E}(f) - E(f, S)]$

2. $\forall \epsilon > 0, \mathbb{P} \left(\sup_{f \in \mathcal{F}} [\mathcal{E}(f) - E(f, S)] - \mathbb{E}_S \sup_{f \in \mathcal{F}} [\mathcal{E}(f) - E(f, S)] > \epsilon \right) \le e^{-2m\epsilon^2}$

3. $\mathbb{E}_S \sup_{f \in \mathcal{F}} (\mathcal{E}(f) - E(f, S)) \le \mathcal{R}_m(L \circ \mathcal{F})$

The first point of the theorem 2.2 is obtained by resolving the equation $e^{-2m\epsilon^2} = \delta$ with respect to ϵ.

3. Bound $\mathcal{R}_m(L \circ \mathcal{F})$ with respect to $R(L \circ \mathcal{F}, S)$

This step is performed by applying the McDiarmid inequality to the function Φ : $S \mapsto R(L \circ \mathcal{F}, S)$. For a given sample S and all $i \in \{1, \ldots, m\}$ consider the set S^i obtained by replacing the example $(\mathbf{x}_i, y_i) \in S$ by $(\mathbf{x}', y')$ generated i. i. d. with respect to the same probability distribution $\mathcal{D}$ generating the examples of S, in this case, the absolute value of the difference $\Phi(S) - \Phi(S^i)$ is bounded by $c_i = 2/m$, and we have from McDiarmid's inequality

$$\forall \epsilon > 0, \mathbb{P}(\mathcal{R}_m(L \circ \mathcal{F}) > R(L \circ \mathcal{F}, S) + \epsilon) \le e^{-m\epsilon^2/2} \qquad (2.32)$$

Thus for $\delta/2 = e^{-m\epsilon^2/2}$, we have with probability at least equal to $1 - \delta/2$:

$$\mathcal{R}_m(L \circ \mathcal{F}) \le R(L \circ \mathcal{F}, S) + 2\sqrt{\frac{\ln \frac{2}{\delta}}{2m}}$$

From the first point (Eq. 2.27) of the theorem 2.2, we have also with probability at least equal to $1 - \delta/2$:

$$\forall f \in \mathcal{F}, \forall S, \mathcal{E}(f) \le E(f, S) + \mathcal{R}_m(L \circ \mathcal{F}) + \sqrt{\frac{\ln \frac{2}{\delta}}{2m}}$$

The second point (Eq. 2.28) of the theorem 2.2 is then obtained by combining the two previous results using the union bound.

As we mentioned at the beginning of this section, the principal advantage of using the Rademacher complexity instead of the VC dimension is that it allows to derive data-dependent generalization error bounds for some usual class of functions and thus have a better estimate of the generalization error over a training set. For example, and in order to illustrate the different steps that are frequently found in the development of the Rademacher complexity, let consider the bounding of the empirical Radmeacher complexity for the class of linear functios with a bounded norm $\mathcal{H}_B = \{h : \mathbf{x} \mapsto \langle \mathbf{x}, \mathbf{w} \rangle \mid ||\mathbf{w}|| \le B\}$ on a given training set S.

$$R(\mathcal{H}_B, S) = \frac{2}{m} \mathbb{E}_\sigma \left[\sup_{f \in \mathcal{H}_B} \left| \sum_{i=1}^{m} \sigma_i h(\mathbf{x}_i) \right| \right]$$

$$= \frac{2}{m} \mathbb{E}_\sigma \left[\sup_{\|\mathbf{w}\| \leq B} \left| \sum_{i=1}^{m} \sigma_i \langle \mathbf{x}_i, \mathbf{w} \rangle \right| \right]$$

From the linearity of the dot product and the Cauchy-Schwarz inequality, we have

$$R(\mathcal{H}_B, S) = \frac{2}{m} \mathbb{E}_\sigma \left[\sup_{\|\mathbf{w}\| \leq B} \left| \left\langle \sum_{i=1}^{m} \sigma_i \mathbf{x}_i, \mathbf{w} \right\rangle \right| \right]$$

$$\leq \frac{2}{m} \mathbb{E}_\sigma \left[\sup_{\|\mathbf{w}\| \leq B} \|\mathbf{w}\| \left\| \sum_{i=1}^{m} \sigma_i \mathbf{x}_i \right\| \right]$$

$$\leq \frac{2B}{m} \mathbb{E}_\sigma \left\| \sum_{i=1}^{m} \sigma_i \mathbf{x}_i \right\|$$

$$= \frac{2B}{m} \mathbb{E}_\sigma \left(\left\langle \sum_{i=1}^{m} \sigma_i \mathbf{x}_i, \sum_{j=1}^{m} \sigma_j \mathbf{x}_j \right\rangle \right)^{1/2}$$

By using the Jensen inequality, and the concavity of the square-root function and the linearity of the dot product again, we have

$$R(\mathcal{H}_B, S) \leq \frac{2B}{m} \left(\mathbb{E}_\sigma \left[\sum_{i=1}^{m} \sum_{j=1}^{m} \sigma_i \sigma_j \langle \mathbf{x}_i, \mathbf{x}_j \rangle \right] \right)^{1/2}$$

The expectation of the double sum over the terms $\sigma_i \sigma_j \langle \mathbf{x}_i, \mathbf{x}_j \rangle$, for $i \neq j$ is equal to zero as in the four possible combinations of the random variables of values -1 and $+1$ and with the same probability $1/2$ there are two with opposite signes, or finally

$$R(\mathcal{H}_B, S) \leq \frac{2B}{m} \sqrt{\sum_{i=1}^{m} \|\mathbf{x}_i\|^2} \tag{2.33}$$

The corollary of this result is a data-dependent generalization bound on the examples of any training set which is obtained for some instantaneous losses and using the properties of the Rademacher complexity that we present in the following

2.3.4 Properties of the Rademacher Complexity

The following states some properties of the Rademacher complexities which are at the basis of most generalization bounds based on this measure.

Theorem 2.4 *Let $\mathcal{F}_1, \ldots, \mathcal{F}_\ell$ and $\mathcal{G}$ be the class of real valued functions. We then have:*

1. *For all $a \in \mathbb{R}$, $R(a\mathcal{F}, S) = |a| R(\mathcal{F}, S)$;*
2. *If $\mathcal{F} \subseteq \mathcal{G}$, then $R(\mathcal{F}, S) \leq R(\mathcal{G}, S)$;*
3. $R\left(\sum_{i=1}^{\ell} \mathcal{F}_i, S\right) \leq \sum_{i=1}^{\ell} R(\mathcal{F}_i, S);$
4. *Let $conv(\mathcal{F})$ be the set of convex combinations (or the convex hull) of $\mathcal{F}$, defined by*

$$conv(\mathcal{F}) = \left\{ \sum_{t=1}^{T} \lambda_t f_t \mid \forall t \in \{1, \ldots, T\}, f_t \in \mathcal{F}, \lambda_t \geq 0 \wedge \sum_{t=1}^{T} \lambda_t = 1 \right\} \quad (2.34)$$

we have then $R(\mathcal{F}, S) = R(conv(\mathcal{F}), S)$;
5. *If $\mathcal{A} : \mathbb{R} \to \mathbb{R}$ is a Lipschitz function with constant L and verifying $\mathcal{A}(0) = 0$ then*

$$R(\mathcal{A} \circ \mathcal{F}, S) \leq 2L R(\mathcal{F}, S);$$

6. *For all function g, $R(\mathcal{F} + g, S) \leq R(\mathcal{F}, S) + \dfrac{2}{m} \left(\sum_{i=1}^{m} g(\mathbf{x}_i)^2 \right)^{1/2}$;*

Properties 1, 2 and 3 derive from the definition of the empirical Rademacher complexity (Eq. 2.20). The condition 5, is at the basis of margin based generalization bounds (Antos et al. 2003) and is most known as Talagrand's lemma (Ledoux and Talagrand 1991, p. 78 corollary 3.17).

Property 6 can be proved by following the same development than the one used for bounding the Rademacher complexity and presented in the previous section. By definition and from the triangular inequality we have

$$\forall g, R(\mathcal{F} + g, S) = \mathbb{E}_\sigma \left[\sup_{f \in \mathcal{F}} \left| \frac{2}{m} \sum_{i=1}^{m} \sigma_i (f(\mathbf{x}_i) + g(\mathbf{x}_i)) \right| \right]$$

$$\leq \frac{2}{m} \mathbb{E}_\sigma \left[\sup_{f \in \mathcal{F}} \left| \sum_{i=1}^{m} \sigma_i f(\mathbf{x}_i) \right| \right] + \frac{2}{m} \mathbb{E}_\sigma \left| \sum_{i=1}^{m} \sigma_i g(\mathbf{x}_i) \right|$$

By using the equality, $\forall z \in \mathbb{R}$, $|z| = \sqrt{z^2}$, we get

$$\forall g, R(\mathcal{F} + g, S) \leq R(\mathcal{F}, S) + \frac{2}{m} \mathbb{E}_\sigma \left[\left(\sum_{i=1}^{m} \sigma_i g(\mathbf{x}_i) \right)^2 \right]^{1/2}$$

$$= R(\mathcal{F}, S) + \frac{2}{m} \mathbb{E}_\sigma \left[\sum_{i=1}^{m} \sum_{j=1}^{m} \sigma_i \sigma_j g(\mathbf{x}_i) g(\mathbf{x}_j) \right]^{1/2}$$

The result then follows from, the Jensen inequality, the concavity of the square root, and the definition of the Rademacher variables, that is:

$$\forall g,\, R(\mathcal{F}+g, S) \le R(\mathcal{F}, S) + \frac{2}{m}\left(\sum_{i=1}^{m} g(\mathbf{x}_i)^2\right)^{1/2}$$

Proof of $R(conv(\mathcal{F}), S) = R(\mathcal{F}, S)$
Let

$$\hat{R}_S(conv(\mathcal{F})) = \frac{2}{m}\mathbb{E}_\sigma\left[\sup_{f_1,\dots,f_T\in\mathcal{F}, \lambda_1,\dots,\lambda_T, \|\lambda\|_1\le 1}\sum_{i=1}^{m}\sigma_i\sum_{t=1}^{T}\lambda_t f_t(\mathbf{x}_i)\right]$$

$$= \frac{2}{m}\mathbb{E}_\sigma\left[\sup_{f_1,\dots,f_T\in\mathcal{F}}\sup_{\lambda_1,\dots,\lambda_T, \|\lambda\|_1\le 1}\sum_{t=1}^{T}\lambda_t\left(\sum_{i=1}^{m}\sigma_i f_t(\mathbf{x}_i)\right)\right]$$

As all the combination weights $\lambda_t, \forall t$ are positive and that $\sum_{t=1}^{T}\lambda_t = 1$ we have $\forall(z_1,\dots,z_T)\in\mathbb{R}^T$:

$$\sup_{\lambda_1,\dots,\lambda_T, \|\lambda\|_1\le 1}\sum_{t=1}^{T}\lambda_t z_t = \max_{t\in\{1,\dots,T\}} z_t$$

That is the supremum is reached by putting all the weights (of sum equal to 1) on the largest term of the combination:

$$\hat{R}_S(conv(\mathcal{F})) = \frac{2}{m}\mathbb{E}_\sigma\left[\sup_{f_1,\dots,f_T\in\mathcal{F}}\sup_{\lambda_1,\dots,\lambda_T, \|\lambda\|_1\le 1}\sum_{t=1}^{T}\lambda_t\left(\sum_{i=1}^{m}\sigma_i f_t(\mathbf{x}_i)\right)\right]$$

$$= \frac{2}{m}\mathbb{E}_\sigma\left[\sup_{f_1,\dots,f_T\in\mathcal{F}}\max_{t\in\{1,\dots,T\}}\sum_{i=1}^{m}\sigma_i f_t(\mathbf{x}_i)\right]$$

$$= \frac{2}{m}\mathbb{E}_\sigma\left[\underbrace{\sup_{h\in\mathcal{F}}\sum_{i=1}^{m}\sigma_i f_t(\mathbf{x}_i)}_{\hat{R}_S(\mathcal{F})}\right] \tag{2.35}$$

Thus, we have

$$R(conv(\mathcal{F}), S) = \max\left(\hat{R}_S(conv(\mathcal{F})), -\hat{R}_S(conv(\mathcal{F}))\right)$$

$$= \max\left(\hat{R}_S(\mathcal{F}), -\hat{R}_S(\mathcal{F})\right)$$

$$= R(\mathcal{F}, S)$$

Chapter 3
Semi-Supervised Learning

Semi-supervised learning, also referred as learning with partially labeled data, concerns the case where a prediction function is learned on both labeled and unlabeled training examples. In this case, labeled examples are usually supposed to be very few leading to an inefficient supervised model, while unlabeled training examples contain valuable information on the prediction problem at hand which exploitation may lead to a performant prediction function. For this scenario, we assume available a set of labeled training examples $S = \{(\mathbf{x}_i, y_i) \mid i = 1, \dots, m\}$ generated from a joint probability distribution $\mathbb{P}(\mathbf{x}, y)$ (also denoted by $\mathcal{D}$) and a set of unlabeled training examples $X_{\mathcal{U}} = \{\mathbf{x}_i \mid i = m+1, \dots, m+u\}$ supposed to be drawn from the marginal distribution $\mathbb{P}(\mathbf{x})$. If $X_{\mathcal{U}}$ is empty, then the problem is cast into the supervised learning framework. The other extreme case corresponds to the situation where S is empty and for which the problem reduces to unsupervised learning. In this chapter we first present basic assumptions (Sect. 3.1) leading to principal approaches developed in semi-supervised learning (Sect. 3.2). We then present two transductive bounds and show their outcomes in terms of sufficient conditions under which unlabeled data may be of help in the learning process (Sect. 3.3.2) and a self-learning algorithm that employs a margin threshold for pseudo-labeling derived from this theoretical study.

3.1 Assumptions

The basic assumption in semi-supervised learning, called *smoothness*, states that

Assumption
Two examples $\mathbf{x}_1$ and $\mathbf{x}_2$, that are close in a high density region should have similar class labels y_1 and y_2.

© Springer International Publishing Switzerland 2015
M.-R. Amini, N. Usunier, *Learning with Partially Labeled and Interdependent Data*,
DOI 10.1007/978-3-319-15726-9_3

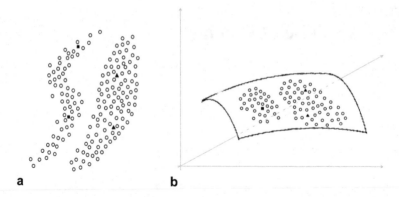

Fig. 3.1 The illustration of cluster and manifold assumptions. Unlabeled examples are represented by *open circles* and labeled examples of different classes are represented by a *filled square* and a *filled triangle*. The partitions of examples are considered as high density regions and the decision boundary should pass through low density regions (cluster assumption, (**a**)). For a given problem, examples are supposed to lie on a geometric manifold of lower dimension (a manifold of dimension 2, or a curved surface on the given example—manifold assumption, (**b**))

This implies that if two points belong to the same group or cluster, so their class labels are likely to be the same. If, on the other hand, they are separated by a low density region, then their desired labels should be different.

Now suppose that the examples of a same class form a partition, the unlabeled training data could then help to find the boundary of each partition more effectively than if only labeled training examples were used. Thus, a way to use unlabeled data to train a model would be to find data partitions using a mixture model and then assign class labels to groups using the labeled data they contain. The underlying assumption, called *cluster assumption*, is therefore

Assumption
If two examples x_1 and x_2 are in the same group, they are likely to belong to the same class y.

This hypothesis could be understood as follows: if there is a group formed by a dense set of examples, then it is unlikely that they belong to different classes. This is not equivalent to say that a class is formed by a single group of examples, but that it is unlikely to find two examples belonging to different classes in the same group. According to the previous smoothness assumption, if we consider the partitions of examples as high density regions, another formulation of cluster assumption, is that the decision boundary passes through low density regions (Fig. 3.1 a).

Density estimation is often based on a notion of distance, which for high dimensional spaces may become meaningless. To overcome this problem a third hypothesis, called *manifold assumption*, on which a number of semi-supervised models are based, states that

> **Assumption**
> In high dimensional spaces, examples lie on low dimensional topological spaces that are locally Euclidian (or geometric manifolds) (Fig. 3.1b).

In the following, we will describe different semi-supervised models that have been developed with respect to the previous assumptions.

3.2 Semi-Supervised Algorithms

In this section we will succinctly present the three main approaches proposed for semi-supervised learning in the literature.

3.2.1 Graphical Approaches

Graphical semi-supervised approaches express the geometry of data by constructing a graph over labeled and unlabeled training examples. The nodes $V = 1, \ldots, m + u$ of this graph G represent the training examples and the edges E reflect the similarity between the examples. These similarities are most often given by a symmetric positive matrix $W = [W_{ij}]_{i,j}$, where $\forall (i, j) \in \{1, \ldots, m + u\}^2$ the weights W_{ij} are nonzero if and only if examples of indexes i and j are connected, or, equivalently, if $(i, j) \in E$. The most used similarity matrices in the literature are:

- the binary matrix of k-nearest neighbors:

$$\forall (i, j) \in \{1, \ldots, m + u\}^2; W_{ij} = 1 \text{ if and only if the example } \mathbf{x}_i \text{ is among the}$$
$$k\text{-nearest neighbours of example } \mathbf{x}_j$$

- the gaussian similarity matrix with parameter σ:

$$\forall (i, j) \in \{1, \ldots, m + u\}^2; W_{ij} = e^{-\frac{\|\mathbf{x}_i - \mathbf{x}_j\|^2}{2\sigma^2}} \tag{3.1}$$

By convention, $W_{ii} = 0$. In the rest of this section we will outline a set of semi-supervised techniques based on the propagation of labels in graphs.

3.2.1.1 Label Propagation

A simple idea to take advantage of G is to propagate the labels of examples throughout the graph. To nodes $1, \dots, m$ associated with the labeled examples are assigned class labels, $+1$ or -1, of these examples, and to nodes of unlabeled examples $m+1, \dots, m+u$ are assigned the label 0. The algorithms following this framework, called label spreading algorithms, are quite similar and they propagate the label of each node of the graph to its neighbors (Zhu and Ghahramani 2002; Zhu et al. 2003; Zhou et al. 2004). The aim of these algorithms is to find the labels $\tilde{Y} = (\tilde{Y}_m, \tilde{Y}_u)$, that are consistent with the class labels of labeled examples, $Y_m = (y_1, \dots, y_m)$, and also with the geometry of examples induced by the structure of the graphe G and expressed by the matrix W.

The consistency between the initial labels of labeled examples, Y_m, and the estimated labels for these examples, $\tilde{Y}_m$, are measured by

$$\sum_{i=1}^{m} (\tilde{y}_i - y_i)^2 = \|\tilde{Y}_m - Y_m\|^2$$

$$= \|S\tilde{Y} - SY\|^2 \tag{3.2}$$

where S is the block diagonal matrix with its m first diagonal elements equal to 1 and its other elements all equal to zero.

The consistency with the geometry of examples, in turn follows the manifold assumption and it consists in penalizing the rapid changes in $\tilde{Y}$ between examples that are close with respect to the matrix W. The latter is thus measured by

$$\frac{1}{2}\sum_{i=1}^{m+u}\sum_{j=1}^{m+u} W_{ij}(\tilde{y}_i - \tilde{y}_j)^2 = \frac{1}{2}\left(2\sum_{i=1}^{m+u}\tilde{y}_i^2\sum_{j=1}^{m+u} W_{ij} - 2\sum_{i,j=1}^{m+u} W_{ij}\tilde{y}_i\tilde{y}_j\right)$$

$$= \tilde{Y}(D \ominus W)\tilde{Y} \tag{3.3}$$

Where $D = [D_{ij}]$ is the diagonal matrix defined by $D_{ii} = \sum_{j=1}^{m+u} W_{ij}$, $\ominus$ represents the matrix substraction and $(D \ominus W)$ is called the unnormalised Laplacian matrix.

The objective function considered here expresses a compromise between two terms (Eq. 3.2 and 3.3):

$$\Delta(\tilde{Y}) = \|S\tilde{Y} - SY\|^2 + \lambda\tilde{Y}(D \ominus W)\tilde{Y} \tag{3.4}$$

Where $\lambda \in (0, 1)$ modulates this compromise. The derivatives of the objective function writes

---

Algorithm 2: Semi-supervised algorithm based on label spreading

---

Input   : Labeled S and unlabeled $X_{\mathcal{U}}$ training sets;
Initialization: $t \leftarrow 0$;
Estimate the similarity matrix $\boldsymbol{W}$ (Eq. 3.1) (For $i \neq j$, $W_{ii} \leftarrow 0$);
Construct the matrix $\boldsymbol{N} \leftarrow \boldsymbol{D}^{-1/2} \boldsymbol{W} \boldsymbol{D}^{-1/2}$ where $\boldsymbol{D}$ is the diagonal matrix defined by
$D_{jj} \leftarrow \sum_j W_{jj}$
Set $\tilde{Y}^{(0)} \leftarrow (y_1, \ldots, y_m, 0, 0, \ldots, 0)$
Choose the parameter $\alpha \in (0,1)$
repeat
    $\tilde{Y}^{(t+1)} \leftarrow \alpha \boldsymbol{N} \, \tilde{Y}^{(t)} + (1-\alpha) \, \tilde{Y}^{(0)}$;
    $t \leftarrow t+1$;
until *convergence*;

Output   : Set $\tilde{Y}^* = (y_1^*, \ldots, y_{m+u}^*)$ the vector obtained after convergence, assign
              class labels to examples $\mathbf{x}_i \in X_{\mathcal{U}}$ following the sign of y_i^*

$$\frac{\partial \Delta(\tilde{Y})}{\partial \tilde{Y}} = 2 \left[\boldsymbol{S}(\tilde{Y} - Y) + \lambda(\boldsymbol{D} \ominus \boldsymbol{W})\tilde{Y} \right]$$

$$= 2 \left[(\boldsymbol{S} \oplus \lambda(\boldsymbol{D} \ominus \boldsymbol{W})) \, \tilde{Y} - \boldsymbol{S}Y \right]$$

Where $\oplus$ represents the matrix addition. Further, the Hessian matrix of the objective function

$$\frac{\partial^2 \Delta(\tilde{Y})}{\partial \tilde{Y} \partial \tilde{Y}^{\top}} = 2 \, (\boldsymbol{S} \oplus \lambda(\boldsymbol{D} \ominus \boldsymbol{W}))$$

is definite positive symmetric, which ensures that the minimum of $\Delta(\tilde{Y})$ is attained when its derivative is equal to zero:

$$\tilde{Y}^* = (\boldsymbol{S} \oplus \lambda(\boldsymbol{D} \ominus \boldsymbol{W}))^{-1} \boldsymbol{S} Y \qquad (3.5)$$

We note that the pseudo-labels of unlabeled examples are obtained by a simple matrix inversion and that this matrix depends only on the unnormalized Laplacian matrix. Other techniques based on the propagation of labels are a variation of the previous formulation. For example the one proposed by (Zhou et al. 2004) is an iterative approach (Algorithm 2) where at each iteration; a node i of the graph receives a contribution from its neighbor j (in the form of a normalized weight of the edge (i, j)) plus a low contribution of its original label, which corresponds to the update rule:

$$\tilde{Y}^{(t+1)} = \alpha N \tilde{Y}^{(t)} + (1 - \alpha)\tilde{Y}^{(0)} \qquad (3.6)$$

Where $\tilde{Y}^{(0)} = (\underbrace{y_1, \ldots, y_m}_{=Y_m}, \underbrace{0, 0, \ldots, 0}_{=Y_u})$ corresponds to the initial vector of labels, $\boldsymbol{D}$

is the diagonal matrix defined by $D_{ii} = \sum_j W_{ij}$, $N = \boldsymbol{D}^{-1/2} \boldsymbol{W} \boldsymbol{D}^{-1/2}$ is the matrix of normalized weights and α is a real value in $(0, 1)$. The proof of convergence of Algorithm 2 follows the update rule (Eq. 3.6). In fact, after t iterations we have

$$\tilde{Y}^{(t+1)} = (\alpha N)^t \tilde{Y}^{(0)} + (1 - \alpha) \sum_{l=0}^{t} (\alpha N)^l \tilde{Y}^{(0)} \tag{3.7}$$

By definition of the diagonal matrix D, the square matrix N is the normalized Laplacian matrix where its elements are positive real values of the interval $[0, 1]$ and the sum of the elements of each line is equal to 1. The matrix N is then by definition a stochastic matrix (or a Markov matrix) and its eigenvalues are less than or equal to 1 (Latouche and Ramaswami 1999, Chap. 2). As α is a real value strictly less than 1, the eigenvalues αN are all strictly less than 1 and we have $\lim_{t \to \infty} (\alpha N)^t = 0$.

Similarly, $\lim_{t \to \infty} \sum_{l=0}^{t} (\alpha N)^l = (I \ominus \alpha N)^{-1}$, where I is the identity matrix. From these results, the vector of pseudo-labels of examples $\tilde{Y}^{(t)}$ converges then to:

$$\lim_{t \to \infty} \tilde{Y}^{(t+1)} = \tilde{Y}^* = (1 - \alpha)(I \ominus \alpha N)^{-1} \tilde{Y}^{(0)} \tag{3.8}$$

The corresponding objective function is:

$$\Delta_n(\tilde{Y}) = \|\tilde{Y} - SY\|^2 + \frac{\lambda}{2} \sum_{i=1}^{m+u} \sum_{j=1}^{m+u} W_{ij} \left(\frac{\tilde{y}_i}{\sqrt{D_{ii}}} - \frac{\tilde{y}_j}{\sqrt{D_{jj}}} \right)^2 \tag{3.9}$$

$$= \|\tilde{Y}_m - Y_m\|^2 + \|\tilde{Y}_u\|^2 + \lambda \tilde{Y}^\top (I \ominus N)\tilde{Y}$$

$$= \|\tilde{Y}_m - Y_m\|^2 + \|\tilde{Y}_u\|^2 + \lambda \left(D^{-1/2}\tilde{Y} \right)^\top (D \ominus W) \left(D^{-1/2}\tilde{Y} \right)$$

In fact, the derivative of this function with respect to pseudo-labels $\tilde{Y}$ is

$$\frac{\partial \Delta_n(\tilde{Y})}{\partial \tilde{Y}} = 2\left[\tilde{Y} - SY + \lambda \left(\tilde{Y} - N\tilde{Y} \right) \right]$$

and it is set to zero for

$$\tilde{Y} = ((1 + \lambda)I \ominus \lambda N)^{-1} SY$$

which for $\lambda = \alpha/(1 - \alpha)$ has the same solution than the one found by Algorithm 2 after convergence (Eq. 3.8). Compared to criterion (Eq. 3.4), the two main differences are (a) the normalized Laplacian matrix and (b) the term $\|\tilde{Y}_m - Y_m\|^2 + \|\tilde{Y}_u\|^2$ that not only forces the pseudo-labels of labeled examples to agree but also the pseudo-labels of unlabeled examples to not reach high values.

3.2.1.2 Markov Random Walks

An alternative to the label spreading algorithm (Algorithm 2), introduced by Szummer and Jaakkola (2002), considers a Markov random walk[1] on the graph G defined with transition probabilities between edges i and j estimated over the similarity matrix by

$$\forall (i, j), p_{ij} = \frac{W_{ij}}{\sum_l W_{il}} \qquad (3.10)$$

The similarity W_{ij} is defined with a gaussian kernel (Eq. 3.1) for the neighbors of the nodes i and j it is fixed to 0 everywhere else. The proposed algorithm first initializes class membership probabilities to $+1$ for all nodes of the graphe, $\mathbb{P}(y = 1 \mid i), i \in V$ using the EM algorithm, and then estimates for each example $\mathbf{x}_j$, the probability $\mathbb{P}^{(t)}(y_s = 1 \mid j)$ to start from an example of class $y_s = 1$ to the example $\mathbf{x}_j$ after t random walks defined as

$$\mathbb{P}^{(t)}(y_s = 1 \mid j) = \sum_{i=1}^{m+u} \mathbb{P}(y = 1 \mid i)\mathbb{P}_{1 \to t}(i \mid j) \qquad (3.11)$$

Where $\mathbb{P}_{1 \to t}(i \mid j)$ is the probability of starting from the node i to arrive to the node j after t random walks. To the node j is finally assigned the pseudo-label of class $+1$, in the case where $\mathbb{P}^{(t)}(y_s = 1 \mid j) > \frac{1}{2}$, and -1 otherwise. In practice, the choice of the value of t has a great impact on the performance of the random walk algorithm and it is not easy to tune. One alternative, proposed by (Zhu and Ghahramani 2002; Zhu et al. 2003) is to assign pseudo-label of class to the node i on the basis of the probability of arriving to the node of class label $+1$ by starting from the node i and arriving to the labeled node, $\mathbb{P}(y_e = 1 \mid i)$, after a random walk. In the case where the example $\mathbf{x}_i$ is labeled, we have

$$\mathbb{P}(y_e = 1 \mid i) = \begin{cases} 1 & \text{if } y_i = 1, \\ 0 & \text{otherwise.} \end{cases}$$

If $\mathbf{x}_i$ is unlabeled, we have the following relation

$$\mathbb{P}(y_e = 1 \mid i) = \sum_{j=1}^{m+u} \mathbb{P}(y_e = 1 \mid j)p_{ij} \qquad (3.12)$$

where p_{ij} is the transition probability defined in (Eq. 3.10). By setting $\forall i, \tilde{z}_i = \mathbb{P}(y_e = 1 \mid i)$ and $\tilde{Z} = (\tilde{Z}_m \ \tilde{Z}_u)$ the corresponding vector divided into two parts

---

[1] A random walk models systems with discrete dynamics composed of a successive random steps (Montroll 1956). The Markov nature of the process reflects the full decorrelation between random steps.

labeled and unlabeled, and by also subdividing the matrices D and W into four parts:

$$D = \begin{pmatrix} D_{mm} & 0 \\ 0 & D_{uu} \end{pmatrix}, \qquad W = \begin{pmatrix} W_{mm} & W_{mu} \\ W_{um} & W_{uu} \end{pmatrix}$$

Equation (3.12) can be written as

$$\tilde{Z}_u = \begin{pmatrix} D_{uu}^{-1} W_{um} & D_{uu}^{-1} W_{uu} \end{pmatrix} \begin{pmatrix} \tilde{Z}_m \\ \tilde{Z}_u \end{pmatrix}$$

$$= D_{uu}^{-1} \left(W_{um} \tilde{Z}_m + W_{uu} \tilde{Z}_u \right) \tag{3.13}$$

Equation (3.13) leads to the following linear system

$$(D \ominus W)_{uu} \tilde{Z}_u = W_{um} \tilde{Z}_m \tag{3.14}$$

We note that if $(\tilde{Z}_m \tilde{Z}_u)$ is a solution to the previous equation, then $(\tilde{Y}_m \tilde{Y}_u)$ defined as

$$\tilde{Y}_m = 2\tilde{Z}_m - 1_m = Y_m$$

$$\tilde{Y}_u = 2\tilde{Z}_u - 1_u$$

where 1_m and 1_u are respectively vectors of dimension m and u. The latter allows to express the linear system with respect to the vectors of labels and pseudo-labels of examples, that is:

$$\tilde{Y}_u = (D \ominus W)_{uu}^{-1} W_{um} Y_m$$

3.2.2 Generative Methods

Learning a prediction function with partially labeled data and using generative models implies the estimation of conditional probabilities $\mathbb{P}(\mathbf{x} \mid y, \Theta)$. As in the unsupervised case; the EM (Dempster et al. 1977) or CEM (Celeux and Govaert 1992) algorithms are often employed to estimate the model parameters Θ. In this case, the hidden variables associated to labeled exemples are known and they correspond to the class labels of these examples and the underlying hypothesis is the cluster assumption (Sect. 3.1), as each partition of unlabeled examples corresponds to a class (Seeger 2001). We can interpret the semi-supervised learning with generative models (a) as a supervised classification where one has additional information on the probability density $\mathbb{P}(\mathbf{x})$ of examples, or (b) as clustering with additional information on the class labels of a subset of examples (Basu et al. 2002). (Machlachlan 1992, p. 39) extended the classification likelihood criterion to semi-supervised learning with generative models. In this case, the class indicator vectors of labeled examples, $(\mathbf{z}_i)_{i=1}^m$, are known and kept fixed during the estimation and classification stages, while group membership vectors of unlabeled examples, $(\tilde{\mathbf{z}}_i)_{i=m+1}^{m+u}$, are estimated. The complete data log likelihood criterion in the semi-supervised case writes:

$$\mathcal{L}_{ssC}(P, \Theta) = \sum_{i=1}^{m} \sum_{k=1}^{K} z_{ik} \ln \mathbb{P}(\mathbf{x}_i, y = k, \Theta) + \sum_{i=m+1}^{m+u} \sum_{k=1}^{K} \tilde{z}_{ik} \ln \mathbb{P}(\mathbf{x}_i, \tilde{y} = k, \Theta)$$

(3.15)

In this expression, the first summation is over labeled examples, and the second one over the unlabeled data. Some work has proposed to modulate the effect of unlabeled examples in the learning stage by adding a real value hyper-parameter $\lambda \in [0, 1]$ to the second term of Eq. 3.15 (Grandvalet and Bengio 2005; Chapelle et al. 2006, Chap. 9). In contrast, the semi-supervised version of the maximum likelihood criterion writes

$$\mathcal{L}_{ssM}(\Theta) = \sum_{i=1}^{m} \ln \mathbb{P}(\mathbf{x}_i, y_i, \Theta) + \sum_{i=m+1}^{m+u} \ln \left(\sum_{k=1}^{K} \pi_k \mathbb{P}(\mathbf{x}_i \mid \tilde{y} = k, \theta_k) \right) \quad (3.16)$$

In the seminal work of Nigam et al. (2000) the parameters of a multinomial naive Bayes classifier for document classification were estimated by maximzing the previous semi-supervised data log-likelihood. The maximization of the complete data log likelihood is however more tractable in practice as it does not involve the logarithm of a sum as in (Eq. 3.16). In the following, we present an extension of the CEM algorithm that lends best to this task. In this case, clusters are assimilated to classes and the initial density components $\mathbb{P}(\mathbf{x} \mid \tilde{y}^{(0)} = k, \theta^{(0)}), k \in \{1, \dots, K\}$ are estimated using labeled examples. At the classification step of the algorithm (C-step), the group membership vectors of unlabeled examples $(\tilde{\mathbf{z}}_i)_{i=m+1}^{m+u}$ are estimated while the class indicator vectors of labeled examples are kept fixed to their known values (Algorithm 3 presents this extension). It is easy to verify that the algorithm converges to a local maxima of (Eq. 3.15). A major difference with the unsupervised CEM algorithm is that after the learning stage the model is used to infer class labels for new examples, while in clustering, the aim is not to build a general inductive rule but just to find the clusters of existing unlabeled examples.

3.2.3 Discriminant Models

The generative approach to semi-supervised learning indirectly computes posteriors $\mathbb{P}(\tilde{y} = k \mid \mathbf{x}, \Theta)$ via conditional density estimation. This is known to lead to poor estimates for high dimensions or when only few data are labeled which is exactly the interesting case for semi-supervised learning (Cohen et al. 2004). A more natural approach would be to use a discriminant model in order to directly estimate posterior probabilities, like the logistic regression model. It can be shown that in this case, the maximization of complete data log-likelihood (Eq. 3.15) is equivalent to the minimization of the empirical classification risk over the labeled and pseudo-labeled training examples. The extension of the semi-supervised CEM algorithm to the discriminant case will be equivalent to the decision-directed, or the self-training algorithm, which consists in using the predicted outputs of the model over unlabeled

---

Algorithm 3: Semi-supervised CEM algorithm

Input   : A labeled training set $S = \{(x_i, y_i) \mid i \in \{1, \ldots, m\}\}$, and an unlabeled training set $X_u = \{x_i \mid i \in \{m+1, \ldots, m+u\}\}$;

Initialization: estimate the parameters $\Theta^{(0)}$ over the labeled examples S;

pour $t \geq 0$ faire

E-step: With the current parameters $\Theta^{(t)} = \{\pi^{(t)}, \theta^{(t)}\}$, estimate the conditional expectations of the group membership vectors of unlabeled examples $\mathbb{E}[\tilde{z}_{ik}^{(t)} \mid x_i; P^{(t)}, \pi^{(t)}, \theta^{(t)}]$. That is $\forall x_i \in X_u$,

$$\pi^{(t)} : \mathbb{E}\left[\tilde{z}_{ik}^{(t)} \mid \text{obs}_i; P^{(t)}, \pi^{(t)}, \theta^{(t)}\right]$$

$$\mathbb{E}\left[\tilde{z}_{ik}^{(t)} \mid x_i; P^{(t)}, \Theta^{(t)}\right] = \mathbb{P}(\tilde{y}^{(t)} = k \mid x_i, \Theta^{(t)}) = \frac{\pi_k^{(t)} \mathbb{P}(x_i \mid \tilde{y}^{(t)} = k, \theta_k^{(t)})}{\sum\limits_{k=1}^{K} \pi_k^{(t)} \mathbb{P}(x_i \mid \tilde{y}^{(t)} = k, \theta_k^{(t)})}$$

C-step: To each unlabeled example $x_i \in X_u$ assign a class using the Bayes decision rule. Denote $P^{(t+1)}$ this new partition.

M-step: Find the new parameters $\Theta^{(t+1)}$ that maximize the complete data log-likelihood (Eq. 3.15);

Output   : The model parameters Θ^*

---

examples to built their class labels (Fralick 1967; Patrick et al. 1970). The process of pseudo-labeling and training a new model is repeated until there are no changes in the pseudo-labels of examples. In the case where the pseudo-labeling is done by thresholding the outputs of the classifier, it can be shown that the self-training algorithm follows the cluster assumption. In the next sections, we will describe the algorithm for binary classification and will present its extension to multiview learning.

Let us rewrite the complete data log-likelihood so as to put in evidence the role of posterior probabilities:

$$\mathcal{L}_{ssC}(P, \Theta) = \sum_{i=1}^{m} \sum_{k=1}^{2} z_{ik} \ln \mathbb{P}(y = k \mid \mathbf{x}_i, \Theta) + \sum_{i=1}^{m} \ln \mathbb{P}(\mathbf{x}_i, \Theta) +$$

$$\sum_{i=m+1}^{m+u} \sum_{k=1}^{2} \tilde{z}_{ik} \ln \mathbb{P}(\tilde{y} = k \mid \mathbf{x}_i, \Theta) + \sum_{i=1}^{m} \ln \mathbb{P}(\mathbf{x}_i, \Theta)$$

As no assumption is made on the distributional nature of data, maximizing $\mathcal{L}_{ssC}$ is equivalent to the maximization of the following criterion (Machlachlan 1992, p. 261):

$$\mathcal{L}_{ssD}(P, \Theta) = \sum_{i=1}^{m} \sum_{k=1}^{2} z_{ik} \ln \mathbb{P}(y = k \mid \mathbf{x}_i, \Theta) + \sum_{i=m+1}^{m+u} \sum_{k=1}^{2} \tilde{z}_{ik} \ln \mathbb{P}(\tilde{y} = k \mid \mathbf{x}_i, \Theta)$$

$$(3.17)$$

In the case where a logistic model estimates the posteriors using a linear function with parameters w: $h_w : \mathcal{X} \to \mathbb{R}$, i.e. $\mathbb{P}(y = 1 \mid \mathbf{x}_i) = \frac{1}{1+e^{-h_w(\mathbf{x}_i)}}$ the maximization of (3.17) is equivalent to the minimization of the semi-supervised surrogate loss

$$\hat{\mathcal{L}}(w) = \frac{1}{m} \sum_{i=1}^{m} \ln\left(1 + e^{-y_i h_w(\mathbf{x}_i)}\right) + \frac{1}{u} \sum_{i=m+1}^{m+u} \ln\left(1 + e^{-\tilde{y}_i h_w(\mathbf{x}_i)}\right) \tag{3.18}$$

that is optimized by the self-training algorithm. A prediction function h_w is first trained on the labeled training set S. The outputs of the current prediction function are then used to estimate the posteriors for unlabeled data. Each unlabeled example $\mathbf{x}_i \in X_{\mathcal{U}}$ is assigned to the class with maximum posterior (**C**-step) and indicator variables are defined accordingly. Using this set of labels, a new prediction function is trained by minimising the semi-supervised surrogate (Eq. 3.18 - **M**-step). These new estimators are then used in the next iteration so as to provide new posterior estimates and therefore new labels for the unlabeled data. Note that in this algorithm, labels for labeled data are always kept fixed to their *true* value since they are known. The **E**-step is trivial here since the posterior estimates are given by the prediction function outputs, it does not explicitly appear in the algorithm, which iterates the two steps C and M until it converges to a local maximum of $\hat{\mathcal{L}}$ (Eq. 3.18). The algorithm is depicted in Fig. 3.2.

The pseudo-labeling of examples that is just based on the outputs of the current prediction function does not guarantee the success of the self-training algorithm compared to the same prediction function trained with only labeled training examples (Chapelle et al. 2006). This is mainly due to the fact that the self-learning algorithm assigning pseudo-class labels without any constraints on the outputs of the learned prediction function, does not follow any of the semi-supervised learning assumptions (Sect. 3.1). To address this problem, some studies modified the classification step by assigning pseudo-labels with respect to a fixed threshold $\rho > 0$, on the outputs of the prediction function and by assuming that the current prediction function does most of its errors on examples that are near the decision boundary, or on low margin examples (Tür et al. 2005):

$$\forall \mathbf{x}_i \in X_{\mathcal{U}}, \tilde{y}_i = \begin{cases} 1 & \text{si, } h(\mathbf{x}_i) \geq \rho \\ -1 & \text{si, } h(\mathbf{x}_i) \leq -\rho \end{cases} \tag{3.19}$$

This hypothesis states that the decision boundary passes through low density regions where the classification errors can be made. With this variant, pseudo-class labels are then assigned to the portion of unlabeled examples that the prediction function is the most confident (with respect to the threshold ρ). This set of examples and there pseudo-labels, noted by $\tilde{S}_{\mathcal{U}}$, is kept fixed during the remaining iterations and both the pseudo-labeling and the the training steps are iterated until either no remaining unlabeled example can be labeled, or all unlabeled examples have been labeled. In this case, the objective function is slightly different than the surrogate loss (Eq. 3.18) and it writes:

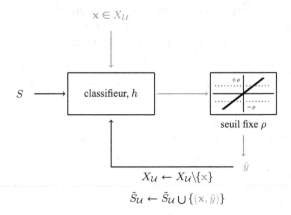

Fig. 3.2 Illustration of the self-training algorithm (Tür et al. 2005). A prediction function $h : \mathcal{X} \to \mathbb{R}$ is first trained over the label training set S. The algorithm then assigns pseudo-label of classes to unlabeled training examples $X_{\mathcal{U}}$ by thresholding the outputs of the prediction function h on these examples (in *grey*). Pseudo-labeled examples are then removed from the set $X_{\mathcal{U}}$ and added to the set of pseudo-labeled examples $\tilde{S}_{\mathcal{U}}$ and a new prediction function is trained using both sets S and $\tilde{S}_{\mathcal{U}}$. These training and pseudo-labeling steps are repeated until convergence

$$\hat{\mathcal{L}}(w) = \frac{1}{m} \sum_{i=1}^{m} \ln\left(1 + e^{-y_i h_w(\mathbf{x}_i)}\right) + \frac{1}{|\tilde{S}_{\mathcal{U}}|} \sum_{\mathbf{x}_i \in \tilde{S}_{\mathcal{U}}} \ln\left(1 + e^{-\tilde{y}_i h_w(\mathbf{x}_i)}\right) \qquad (3.20)$$

In the following section we assume that a classifier $f : \mathcal{X} \to \{-1, +1\}$, is obtained on the basis of a learned prediction function $h : \mathcal{X} \to \mathbb{R}$ using the sign function:

$$\forall \mathbf{x} \in \mathcal{X}, \ f(\mathbf{x}) = \text{sgn}(h(\mathbf{x}))$$

3.3 Transductive Learning

Transductive learning, introduced by Vapnik (1999), tends to produce a prediction function for only a fixed number of unlabeled examples $X_{\mathcal{U}} = \{\mathbf{x}_i \mid i = m+1, \ldots, m+u\}$. The framework is motivated by the fact that for some applications it is not necessary to learn a general rule as in the inductive case, but just to accurately predict the outputs of examples of an unlabeled or a test set[2]. The transductive error defined by this type of learning is then the average number of predicted outputs $\{\tilde{y}_i \mid \mathbf{x}_i \in X_{\mathcal{U}}, i \in \{m+1, \ldots, m+u\}\}$ different from the true class labels

---

[2] The transductive learning is a special case of semi-supervised learning, as the unlabeled examples are known a priori and are observed by the learning algorithm during the training stage.

of unlabeled examples $\{y_i \mid \mathbf{x}_i \in X_\mathcal{U}, i \in \{m+1, \dots, m+u\}\}$:

$$R_u(\mathcal{T}) = \frac{1}{u} \sum_{i=m+1}^{m+u} \mathbb{1}_{[y_i \neq \tilde{y}_i]} \tag{3.21}$$

Where $\mathcal{T} : \mathcal{X} \rightarrow \mathcal{Y}$ is the transductive algorithm that predicts the output $\tilde{y}_i \in \mathcal{Y}$ for the unlabeled example $\mathbf{x}_i \in X_\mathcal{U}$. As the size of this unlabeled set is finite, the considered function class, $\mathcal{F} = \{-1, +1\}^{m+u}$, for finding the transductive prediction function $\mathcal{T}$ is also finite (Derbeko et al. 2003). According to the structural risk minimization principle (SRM—Chap. 2, Sect. 2.2.3), $\mathcal{F}$ can then be defined according to a nested structure

$$\mathcal{F}_1 \subset \mathcal{F}_2 \subset \dots \subset \mathcal{F} = \{-1, +1\}^{m+u} \tag{3.22}$$

3.3.1   Transductive Support Vector Machines

This structure normally reflects a priori knowledge on the learning problem at hand and must be constructed in the way that with high probability, the correct prediction of class labels of labeled and unlabeled training examples is contained in a function class $\mathcal{F}_k$ of small size. In particular Derbeko et al. (2003) shows that for any $\delta \in (0, 1)$ we have

$$\mathbb{P}\left(R_u(\mathcal{T}) \leq R_m(\mathcal{T}) + \Gamma(m, u, |\mathcal{F}_k|, \delta)\right) \geq 1 - \delta \tag{3.23}$$

Where $R_m(\mathcal{T})$ is the transductive error (or the empirical classification loss) on a training set $\mathcal{T}$ and $\Gamma(m, u, |\mathcal{F}_k|, \delta)$ is a complexity term depending on the number of labeled examples m, the number of unlabeled examples u, and the size $|\mathcal{F}_k|$ of the function class $\mathcal{F}_k$. To find the best function class among the existing ones, transductive algorithms generally use the distribution of unsigned margins of unlabeled examples in order to guide the search of a prediction function by constraining it to respect semi-supervised learning assumptions. According to the SRM principle, the learning algorithm should choose the pseudo-labels of unlabeled examples $\{\tilde{y}_i \mid \mathbf{x}_i \in X_\mathcal{U}\} \in \mathcal{F}_k$ for which the empirical loss $R_m(\mathcal{T})$ and the size of the function class $|\mathcal{F}_k|$ minimize the upper bound (3.23). In the particular case where the empirical loss is zero, the minimization of the upper bound is equivalent to find the labels of unlabeled examples with the largest margin.

Transductive Support Vector Machine (TSVM) are derived from this paradigm, and they find the best hyperplane in a feature space that separates the best labeled exemples and that does not pass through high density regions. For this, TSVM constructs a structure on a function class $\mathcal{F}$ by ordering the outputs of unlabeled examples with respect to their margins. The corresponding optimization problem, known as hard margin TSVM writes (Vapnik 1999):

$$\min_{\bar{w}, w_0, \tilde{y}_{m+1}, \dots, \tilde{y}_{m+u}} \frac{1}{2} ||\bar{w}||^2 \tag{3.24}$$

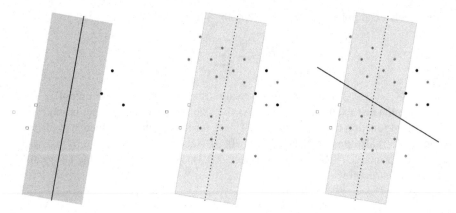

Fig. 3.3 Illustration of hyperplanes found by the SVM algorithm (*left*) and the TSVM algorithm (*right*) for a binary classification problem, where labeled examples are represented by *circles* and *squares* while unlabeled examples are represented by *stars*. The SVM algorithm finds the decision boundary by ignoring unlabeled examples while the TSVM algorithm find the hyperplane that separates labeled examples with high margin without crossing dense regions

$$\text{u.c. } \forall i \in \{1, \ldots, m\}, y_i \left(\langle \bar{w}, \mathbf{x}_i \rangle + w_0 \right) - 1 \geq 0$$

$$\forall i \in \{m+1, \ldots, m+u\}, \tilde{y}_i \left(\langle \bar{w}, \mathbf{x}_i \rangle + w_0 \right) - 1 \geq 0$$

$$\forall i \in \{m+1, \ldots, m+u\}, \tilde{y}_i \in \{-1, +1\} \tag{3.25}$$

The solutions to this optimization problem are the pseudo-labels of unlabeled examples $\tilde{y}_{m+1}, \ldots, \tilde{y}_{m+u}$ for which the hyperplane of parameters $\{\bar{w}, w_0\}$ separates the examples of both labeled and unlabeled examples with the largest margin. Figure 3.3 illustrates the solutions found by the SVM and the TSVM algorithms on a toy problem. The main difference between these two algorithms is that the hyperplane found by SVM is the one that has the largest margin on labeled examples regardless of the position of unlabeled examples. By introducing slack variables associated to labeled and unlabeled examples, Joachmis (1999) extended the optimization problem 3.24 to the non-separable case

$$\min_{\bar{w}, w_0, \tilde{y}_{m+1}, \ldots, \tilde{y}_{m+u}, \xi_1, \ldots, \xi_m, \tilde{\xi}_1, \ldots, \tilde{\xi}_u} \frac{1}{2} ||\bar{w}||^2 + C \sum_{i=1}^{m} \xi_i + \tilde{C} \sum_{i=1}^{u} \tilde{\xi}_i \tag{3.26}$$

$$\text{u.c.} \forall i \in \{1, \ldots, m\}, y_i \left(\langle \bar{w}, \mathbf{x}_i \rangle + w_0 \right) \geq 1 - \xi_i$$

$$\forall i \in \{m+1, \ldots, m+u\}, \tilde{y}_i \left(\langle \bar{w}, \mathbf{x}_i \rangle + w_0 \right) \geq 1 - \tilde{\xi}_{i-m}$$

$$\forall i \in \{m+1, \ldots, m+u\}, \tilde{y}_i \in \{-1, +1\} \tag{3.27}$$

$$\forall i \in \{1, \ldots, m\}, \xi_i \geq 0$$

$$\forall i \in \{1, \ldots, u\}, \tilde{\xi}_i \geq 0$$

For both problems (3.24) and (3.26), the pseudo-labels of unlabeled examples are integers and according to the constraints (3.25) and (3.27), these two problems write as an optimization problem with quadratic objective functions with linear constraints. Failing to have an efficient solver at the moment, some work proposed an approximate solution to this problem by employing for example the coordinate descent technique (Joachmis 1999, 2002a) which consists in choosing alternatively a coordinate and to minimize the objective function on the chosen direction with simple techniques like the line search algorithm.

3.3.2 A Transductive Bound for the Voted Classifier

In this section, we pressent a bound over the transductive risk (Eq. 3.21) of a Bayes classifier by considering the distribution of unsigned margins of this classifier over unlabeled training examples. The expression of a Bayes classifier, defined with a probability distribution Q over a function class $\mathcal{F}$, is:

$$B_Q : \mathcal{X} \to \{-1, +1\}$$
$$\mathbf{x} \mapsto \mathrm{sgn}\left[\mathbb{E}_{f \sim Q} f(\mathbf{x})\right] \tag{3.28}$$

B_Q translates the majority vote of classifiers in $\mathcal{F}$ with their respective weights, over the class labels of examples. The transductive risk of the Bayes classifier over the unlabeled training examples $X_\mathcal{U}$ (Eq. 3.21), writes then

$$R_u(B_Q) = \frac{1}{u} \sum_{\mathbf{x} \in X_\mathcal{U}} \mathbb{1}_{[B_Q(\mathbf{x}) \neq y]} \tag{3.29}$$

Similarly, we define the associated Gibbs classifier, G_Q, as the prediction function obtained with random sampling according to the distribution from $\mathcal{F}$ and its tranductive risk is:

$$R_u(G_Q) = \frac{1}{u} \sum_{\mathbf{x} \in X_\mathcal{U}} \mathbb{E}_{f \sim Q}\left[\mathbb{1}_{[f(\mathbf{x}) \neq y]}\right] \tag{3.30}$$

By considering the unsigned margin distribution of unlabeled examples

$$\forall \mathbf{x} \in \mathcal{X}, m_Q(\mathbf{x}) = \left|\mathbb{E}_{f \sim Q} f(\mathbf{x})\right| \tag{3.31}$$

We can bound the transductive risk of the Bayes classifier on examples having an unsigned margin above a fixed threshold ρ as:

$$R_{u \wedge \rho}(B_Q) = \frac{1}{u} \sum_{\mathbf{x} \in X_\mathcal{U}} \mathbb{1}_{[B_Q(\mathbf{x}) \neq y \wedge m_Q(\mathbf{x}) > \rho]} \tag{3.32}$$

Theorem 3.1 (*Amini et al. 2009*). *Let B_Q be the Bayes classifier defined as in (Eq. 3.28), then for all Q, all $\rho \geq 0$ and all $\delta \in (0, 1]$ the following inequality holds with probability at least $1 - \delta$:*

$$R_{u \wedge \rho}(B_Q) \leq \inf_{\gamma \in (\rho, 1]} \left\{ \mathbb{P}_u(\rho < m_Q(\mathbf{x}) < \gamma) + \frac{1}{\gamma} \left\lfloor K_u^\delta(Q) + M_Q^{\leq}(\rho) - M_Q^{\leq}(\gamma) \right\rfloor_+ \right\} \tag{3.33}$$

Where $\mathbb{P}_u(C)$ is the fraction of unlabeled examples verifying the condition C, $K_u^\delta(Q) = R_u^\delta(G_Q) + \frac{1}{2}(\mathbb{E}_u[m_Q(\mathbf{x})] - 1)$, $M_Q^\triangleleft(z) = \mathbb{E}_u\left[m_Q(\mathbf{x})\mathbb{1}_{[m_Q(\mathbf{x})\triangleleft z]}\right]$ with $\triangleleft$ being $<$ or $\leq$ and $\lfloor z \rfloor_+ = \mathbb{1}_{[z>0]}z$

PROOF **of Theorem 3.1**

1. Consider the set of unsigned margins of unlabeled examples of size N:

$$\{\gamma_i, i = 1, \ldots, N\} = \{m_Q(\mathbf{x}) \mid \mathbf{x} \in X_U \wedge m_Q(\mathbf{x}) > 0\}$$

Let $k = \max\{i \mid \gamma_i \leq \rho\}$, the largest index of margins smaller than ρ, and $b_i = \mathbb{P}_u(B_Q(\mathbf{x}) \neq y \wedge m_Q(\mathbf{x}) = \gamma_i), \forall i \in \{1, \ldots, N\}$, the fraction of unlabeled examples misclassified by the Bayes classifier and having unsigned margins is equal to γ_i. The transductive risk of the Bayes classifier over unlabeled examples having an unsigned margin above a given threshold ρ (3.32) can be written as

$$\forall \rho \in [0, 1], R_{u\wedge\rho}(B_Q) = \sum_{i=k+1}^{N} b_i \tag{3.34}$$

2. By the definition of unsigned margins and considering the fact that functions $f \in \mathcal{F}$ have values in $\{-1, +1\}$ we have

$$\forall \mathbf{x} \in X_U, m_Q(\mathbf{x}) = |\mathbb{E}_{f\sim Q}f(\mathbf{x})| = |1 - 2\mathbb{E}_{f\sim Q}[\mathbb{1}_{[f(\mathbf{x})\neq y]}]| \tag{3.35}$$

Moreover, the Bayes classifier misclassifies an example $\mathbf{x} \in X_U$ if and only if $\mathbb{E}_{f\sim Q}\mathbb{1}_{[f(\mathbf{x})\neq y]} > \frac{1}{2}$ which according to Eq. (3.35) gives:

$$\mathbb{E}_{f\sim Q}[\mathbb{1}_{[f(\mathbf{x})\neq y]}] = \frac{1}{2}(1 + m_Q(\mathbf{x}))\mathbb{1}_{[B_Q(\mathbf{x})\neq y]} + \frac{1}{2}(1 - m_Q(\mathbf{x}))\mathbb{1}_{[B_Q(\mathbf{x})=y]} \tag{3.36}$$

By averaging the previous equality over all unlabeled examples, the risk of the Gibbs classifier (3.30) can be expressed according to b_i and γ_i

$$R_u(G_Q) = \frac{1}{u}\sum_{\mathbf{x}\in X_U} m_Q(\mathbf{x})\mathbb{1}_{[B_Q(\mathbf{x})\neq y]} + \frac{1}{2}(1 - \mathbb{E}_u[m_Q(\mathbf{x})])$$

$$= \sum_{i=1}^{N} b_i\gamma_i + \frac{1}{2}(1 - \mathbb{E}_u[m_Q(\mathbf{x})]) \tag{3.37}$$

3. Now suppose that we have a bound $R_u^\delta(G_Q)$ on the risk of the Gibbs classifier, $R_u(G_Q)$ that holds with probability at least $1 - \delta$, according to Eqs. (3.34) and (3.37), we have in the worst case a bound on the joint risk of the Bayes classifier:

$$R_{u \wedge \rho}(B_Q) \leq \max_{b_1,\ldots,b_N} \sum_{i=k+1}^{N} b_i$$

$$\text{u.c. } \forall i, 0 \leq b_i \leq \mathbb{P}_u(m_Q(\mathbf{x}) = \gamma_i) \text{ and } \underbrace{\sum_{i=1}^{N} b_i \gamma_i \leq R_u^\delta(G_Q) - \frac{1}{2}(1 - \mathbb{E}_u[m_Q(\mathbf{x})])}_{K_u^\delta(Q)}$$

This corresponds to a linear optimization problem (Dantzig 1951) and the solution which maximizes $\sum_{i=k+1}^{N} b_i$ is a convex polyhedron given by

$$b_i = \begin{cases} 0 & \text{if } i \leq k \\ \min\left(\mathbb{P}_u(m_Q(\mathbf{x}) = \gamma_i), \left\lfloor \frac{K_u^\delta(Q) - \sum_{k<j<i} \gamma_j \mathbb{P}_u(m_Q(\mathbf{x})=\gamma_j)}{\gamma_i} \right\rfloor_+ \right) & \text{otherwise} \end{cases}$$

Let $I = \max\left\{ i \mid K_u^\delta(Q) - \sum_{k<j<i} \gamma_j \mathbb{P}_u(m_Q(\mathbf{x}) = \gamma_j) > 0 \right\}$, we have then:

- If $i < I, b_i = \mathbb{P}_u(m_Q(\mathbf{x}) = \gamma_i)$,
- If $i = I, b_I = \frac{K_u^\delta(Q) - \sum_{k<j<I} \gamma_j \mathbb{P}_u(m_Q(\mathbf{x})=\gamma_j)}{\gamma_I}$,
- If $i > I, b_i = 0$

That is

$$R_{u \wedge \rho}(B_Q) \leq \mathbb{P}_u(\rho < m_Q(\mathbf{x}) < \gamma_I) + \frac{K_u^\delta(Q) + M_Q^\leq(\rho) - M_Q^\leq(\gamma_I)}{\gamma_I} \quad (3.38)$$

Where $M_Q^\leq(\gamma_i) = \sum_{j=1}^{j<i} \gamma_j \mathbb{P}_u(m_Q(\mathbf{x}) = \gamma_j)$. The proof is ended by noting that the following function is decreasing function for $\gamma < \gamma_I$

$$\Psi : \gamma \mapsto \mathbb{P}_u(\theta < m_Q(\mathbf{x}) < \gamma) + \frac{K_u^\delta(Q) + M_Q^\leq(\rho) - M_Q^\leq(\gamma)}{\gamma}$$

In the case where $\rho = 0$ and $M_Q^\leq(0) = 0$ and by definition

$$R_u(B_Q) = R_{u \wedge 0}(B_Q) + \mathbb{P}_u(B_Q(\mathbf{x}) \neq y \wedge m_Q(\mathbf{x}) = 0) \leq R_{u \wedge 0}(B_Q) + \mathbb{P}_u(m_Q(\mathbf{x}) = 0)$$

we have the following corollary that gives a bound over the transductive risk of the Bayes classifier (Eq. 3.29).

Corollary 3.1 *Let B_Q be the Bayes classifier defined as (Eq. 3.28), then for all Q, all $\delta \in (0, 1]$ we have with probability at least $1 - \delta$:*

$$R_u(B_Q) \leq \inf_{\gamma \in (\rho, 1]} \left\{ \mathbb{P}_u(m_Q(\mathbf{x}) < \gamma) + \frac{1}{\gamma} \lfloor K_u^\delta(Q) - M_Q^<(\gamma) \rfloor_+ \right\} \tag{3.39}$$

$K_u^\delta(Q) = R_u^\delta(G_Q) + \frac{1}{2}(\mathbb{E}_u [m_Q(\mathbf{x})] - 1), M_Q^<(z) = \mathbb{E}_u \left[m_Q(\mathbf{x}) \mathbb{1}_{[m_Q(\mathbf{x})<z]} \right]$ and $\lfloor z \rfloor_+ = \mathbb{1}_{[z>0]} z$

An interesting result of this bound is that in the case where we consider the margin as an indicator of confidence, it is then possible to have a good estimation of the transductive risk of the Bayes classifier following the cluster assumption evoked in Sect. 3.1. This result is stated in the following proposition.

Proposition 3.1 *Suppose that $\forall \mathbf{x} \in X_U, m_Q(\mathbf{x}) > 0, \delta \in (0, 1]$ and $\exists D \in (0, 1]$ such that $\forall \gamma > 0$:*

$$\mathbb{P}_u(B_Q(\mathbf{x}) \neq y \wedge m_Q(\mathbf{x}) = \gamma)$$
$$\neq 0 \Rightarrow \mathbb{P}_u(B_Q(\mathbf{x}) \neq y \wedge m_Q(\mathbf{x}) < \gamma) \geq D\mathbb{P}_u(m_Q(\mathbf{x}) < \gamma)$$

Then with probability at least $1 - \delta$ we have:

$$F_u^\delta(Q) - R_u(B_Q) \leq \frac{1-D}{D} R_u(B_Q) + \frac{R_u^\delta(G_Q) - R_u(G_Q)}{\gamma^*} \tag{3.40}$$

Where $\gamma^* = \sup\{\gamma \mid \mathbb{P}_u(B_Q(\mathbf{x}) \neq y \wedge m_Q(\mathbf{x}) = \gamma) \neq 0\}$ and $F_u^\delta(Q) = \inf_{\gamma \in (\rho, 1]} \left\{ \mathbb{P}_u(m_Q(\mathbf{x}) < \gamma) + \frac{1}{\gamma} \lfloor K_u^\delta(Q) - M_Q^<(\gamma) \rfloor_+ \right\}$ is the transductive bound over the Bayes risk (Eq. 3.39).

PROOF **of Proposition 3.1**
Suppose that

$$R_u(B_Q) \geq \mathbb{P}_u(B_Q(\mathbf{x}) \neq y \wedge m_Q(\mathbf{x}) < \gamma^*) + \frac{1}{\gamma^*} \lfloor K_u(Q) - M_Q^<(\gamma^*) \rfloor_+ \tag{3.41}$$

where γ^* is the largest margin for which the classifier makes an error; $\gamma^* = \sup\{\gamma \mid \mathbb{P}_u(B_Q(\mathbf{x}) \neq y \wedge m_Q(\mathbf{x}) = \gamma) \neq 0\}$ and $K_u(Q) = R_u(G_Q) + \frac{1}{2} (\mathbb{E}_u [m_Q(\mathbf{x})] - 1)$. Considering the assumption, we then have

$$R_u(B_Q) \geq D\mathbb{P}_u(B_Q(\mathbf{x}) < \gamma^*) + \frac{1}{\gamma^*} \lfloor K_u(Q) - M_Q^<(\gamma^*) \rfloor_+ \tag{3.42}$$

For a given $\delta \in (0,1]$, the inequality $F_u^\delta(Q) \leq \mathbb{P}_u(B_Q(\mathbf{x}) < \gamma^*) + \frac{1}{\gamma^*}\lfloor K_u(Q) - M_Q^{\leq}(\gamma^*)\rfloor_+$ holds with probability at least $1 - \delta$, and with the same probability we have

$$F_u^\delta(Q) - R_u(B_Q) \leq (1 - D)\mathbb{P}_u(m_Q(\mathbf{x}) < \gamma^*) + \frac{R_u^\delta(G_Q) - R_u(G_Q)}{\gamma^*} \quad (3.43)$$

This is because $\lfloor K_u^\delta(Q) - M_Q^{\leq}(\gamma^*)\rfloor_+ - \lfloor K_u(Q) - M_Q^{\leq}(\gamma^*)\rfloor_+ \leq R_u^\delta(G_Q) - R_u(G_Q)$. From the inequality (3.42) we also have $\mathbb{P}_u(B_Q(\mathbf{x}) < \gamma^*) \leq \frac{1}{D}R_u(B_Q)$, which by plugging it in equation (Eq. 3.42) gives the result.

The interpretation of the bound (Eq. 3.40) is that following the cluster assumption, the hyperplane induced by the Bayes classifier does not pass through high density regions and it thus makes the most of its errors on low density regions. In this case, the constant D will be near 1, and we can have a very accurate estimation of the error of the Bayes classifier.

3.4 Multiview Learning Based on Pseudo-Labeling

For applications where data are produced by several information sources, like images described by different visual and textual descriptors or multilingual documents wrote on different languages; the pseudo-labeling of examples can be carried out by aggregating the predictions provided by each view specific classifiers. Each representation of an example corresponds to the characterisation of the information provided by an associated source and the paradigm that is based on the exploitation of these views to learn the classifiers is called multiview learning. Formally, a *multiview observation* is a sequence $\mathbf{x} \overset{def}{=} (x^1, ..., x^V)$, where different *views* x^v provide a representation of the same object in different sets $\mathcal{X}_v$.

There exist three families of approaches developed according to this concept and which use the redundancy in different representations of data, by aiming to decrease the disagreement between the view-specific class predictions to increase the overall performance. These algorithms achieve this goal, either by projecting the views specific representations in a common canonical space (Bach et al. 2004), or by constraining the classifiers to have similar outputs on the same examples by adding a disagreement term in their objective functions (Sindhwani et al. 2005), or even by pseudo-labeling unlabeled examples according to the output of view-specific classifiers (Blum and Mitchell 1998a).

The pseudo-labeling approach was initiated by the seminal work of Blum and Mitchell (1998a) who proposed the co-training algorithm for multiview problems with two view. This model assumes that each representation is rich enough to learn the parameters of the corresponding classifier in the case where there are enough labeled examples available. Both classifiers are first trained separately on the labeled data. A subset of unlabeled examples is then randomly drawn and pseudo-labeled by each of the classifiers. The estimated output by the first classifier becomes the desired output for the second classifier and reciprocally. The proposed implementation is quite simple and it is similar to the decision directed algorithm with the difference that there are two classifiers, each alternatively supplying pseudo-labels to the other classifier. Blum and Mitchell (1998a) use this algorithm to train Naive-Bayes classifiers to classify web pages into two relevant and irrelevant classes, by considering two different representations for each page obtained from the terms of links pointing to that page and also from the terms contained in the page.

This algorithm has been shown to be efficient in the context of web page classification but it presents a number of limitations. The most important is, the pseudo-labeling of a part of the unlabeled data that forces both classifiers to slowly agree on their outputs. However, different studies proved that the agreement over the outputs of different view-specific classifiers is the basis of the multiview learning paradigm (Sindhwani et al. 2005; Leskes 2005a). In particular, Leskes (2005a) showed that by constraining view-specific classifiers to have similar outputs on unlabeled data, it is possible to restrict the research in the function class space and hence to have a better estimation of the generalization error of each view-specific classifier by reducing the complexity term involved in their generalization bounds.

3.4.1 Learning with Partially Observed Multiview Data

Amini et al. (2010) considered binary classification problems where, given a multiview observation, some of the views are not observed. This happens, for instance, when documents may be available in different languages, yet a given document may only be available in a single language. Formally, observations $\mathbf{x}$ belong to the input set $\mathcal{X} \stackrel{def}{=} (\mathcal{X}_1 \cup \{\bot\}) \times \ldots \times (\mathcal{X}_V \cup \{\bot\})$, where $x^v = \bot$ means that the v-th view is not observed. In such case, examples are assumed to be pairs $(\mathbf{x}, y)$, with $y \in \mathcal{Y} \stackrel{def}{=} \{0, 1\}$, drawn according to a fixed, but unknown distribution $\mathcal{D}$ over $\mathcal{X} \times \mathcal{Y}$, such that $\mathbb{P}_{(\mathbf{x},y) \sim \mathcal{D}} (\forall v : x^v = \bot) = 0$ (at least one view is available). In multilingual text classification, a *parallel corpus* is a dataset where all views are always observed (i.e. $\mathbb{P}_{(\mathbf{x},y) \sim \mathcal{D}} (\exists v : x^v = \bot) = 0$), while a *comparable corpus* is a dataset where only one view is available for each example (i.e. $\mathbb{P}_{(\mathbf{x},y) \sim \mathcal{D}} (|\{v : x^v \neq \bot\}| \neq 1) = 0$).

For a given observation $\mathbf{x}$, the views v such that $x^v \neq \bot$ are called the *observed views* and it is assumed that there exist *view generating functions* $\Psi_{v \to v'} : \mathcal{X}_v \to \mathcal{X}_{v'}$ which take as input a given view x^v and output an element of $\mathcal{X}_{v'}$, that is supposed to be *close* to what $x^{v'}$ would be if it was observed. In the multilingual text classification example, the view generating functions would be Machine Translation systems. These generating functions can then be used to create surrogate observations, such that all views are available. For a given partially observed $\mathbf{x}$, the *completed* observation $\mathbf{x}$ is obtained as:

$$\forall v, \underline{x}^v = \begin{cases} x^v & \text{if } x^v \neq \perp \\ \Psi_{v' \rightarrow v}(x^{v'}) & \text{otherwise, where } v' \text{ is such that } x^{v'} \neq \perp \end{cases} \tag{3.44}$$

The addressed learning task is to find, in some predefined classifier set $\mathcal{C}$, the stochastic classifier c that minimizes the classification error on multiview examples (with, potentially, unobserved views) drawn according to some distribution $\mathcal{D}$ as described above. Following the standard multiview framework, in which all views are observed (Blum and Mitchell 1998b; Muslea 2002), it is assumed that there are V *deterministic* classifier sets $(\mathcal{F}_v)_{v=1}^V$, each working on one specific view. That is, for each view v, $\mathcal{F}_v$ is a set of functions $f_v : \mathcal{X}_v \rightarrow \{0, 1\}$, where each view specific function is obtained from its corresponding prediction function $h_v : \mathcal{X}_v \rightarrow \mathbb{R}$ using the rule

$$\forall v, \forall x^v \in \mathcal{X}_v, f_v = \frac{1}{2}(\text{sgn}(h_v) + 1)$$

The final set of classifiers $\mathcal{C}$ contains *stochastic* classifiers, whose output only depends on the outputs of the view-specific classifiers. That is, associated to a set of classifiers $\mathcal{C}$, there is a function $\Phi_{\mathcal{C}} : (\mathcal{F}_v)_{v=1}^V \times \mathcal{X} \rightarrow [0, 1]$ such that:

$$\mathcal{C} = \{\mathbf{x} \mapsto \Phi_{\mathcal{C}}(f_1, ..., f_V, \mathbf{x}) \,|\, \forall v, f_v \in \mathcal{F}_v\}$$

The overall objective of learning is therefore to find $c \in \mathcal{C}$ with low generalization error, defined as:

$$\mathcal{E}(c) = \mathbb{E}_{(\mathbf{x},y) \sim \mathcal{D}} [e(c(\mathbf{x}), y)] \tag{3.45}$$

where e is a pointwise error, for instance the $0/1$ loss: $e(c(\mathbf{x}), y) = c(\mathbf{x})(1 - y) + (1 - c(\mathbf{x}))y$.

In the following sections, two learning tasks in terms of supervised and semi-supervised learning are considered.

3.4.1.1 Supervised Learning Tasks

Assume that a training set S of m examples is drawn i.i.d. according to a distribution $\mathcal{D}$, as presented in the previous section. Depending on how the generated views are used at both training and test stages, consider the following learning scenarios:

- Baseline: This setting corresponds to the case where each view-specific classifier is trained using the corresponding observed view on the training set, and prediction for a test example is done using the view-specific classifier corresponding to the observed view:

$$\forall v, f_v \in \underset{f \in \mathcal{F}_v}{\text{argmin}} \sum_{(\mathbf{x},y) \in S : x^v \neq \perp} e(f(x^v), y) \tag{3.46}$$

In this case $\forall \mathbf{x}, c^b_{f_1,...,f_V}(\mathbf{x}) = f_v(x^v)$, where v is the observed view for $\mathbf{x}$. Notice that this is the most basic way of learning a text classifier from a comparable corpus.

- Generated Views as Additional Training Data: The most natural way to use the generated views for learning is to use them as additional training material for the view-specific classifiers:

$$\forall v, f_v \in \underset{f \in \mathcal{F}_v}{\text{argmin}} \sum_{(\mathbf{x},y) \in S} e(f(\underline{x}^v), y) \qquad (3.47)$$

with $\underline{x}$ defined by Eq. 3.44. Prediction is still done using the view-specific classifiers corresponding to the observed view, i.e. $\forall \mathbf{x}, c^b_{f_1,...,f_V}(\mathbf{x}) = f_v(x^v)$. Although the test set distribution is a subdomain of the training set distribution (Blitzer et al. 2007), this mismatch is (hopefully) compensated by the addition of new examples.

- multiview Gibbs Classifier: In order to avoid the potential bias introduced by the use of generated views only during training, these generated views are also considered during testing. This becomes a standard multiview setting, where generated views are used exactly as if they were observed. The view-specific classifiers are trained exactly as above (Eq. 3.47), but the prediction is carried out with respect to the probability distribution of classes, by estimating the probability of class membership in class 1 from the mean prediction of each view-specific classifier:

$$\forall \mathbf{x}, c^{mg}_{f_1,...,f_V}(\mathbf{x}) = \frac{1}{V} \sum_{v=1}^{V} f_v(\underline{x}^v) \qquad (3.48)$$

- multiview Majority Voting: With view generating functions involved in training and test, a natural way to obtain a (generally) deterministic classifier with improved performance is to take the majority vote associated with the Gibbs classifier. The view-specific classifiers are again trained as in Eq. 3.47, but the final prediction is done using a majority vote:

$$\forall \mathbf{x}, c^{mv}_{f_1,...,f_V}(\mathbf{x}) = \begin{cases} \frac{1}{2} & \text{if } \sum_{v=1}^{V} f_v(\underline{x}^v) = \frac{V}{2} \\ \mathbb{1}_{\left[\sum_{v=1}^{V} f_v(\underline{x}^v) > \frac{V}{2}\right]} & \text{otherwise} \end{cases} \qquad (3.49)$$

The classifier outputs either the majority voted class, or either one of the classes with probability $1/2$ in case of a tie.

3.4.1.2 The Trade-Offs with the ERM Principle

The following theorem provides bounds on the baseline vs. multiview strategies. The notion of function class capacity used here is the *empirical Rademacher complexity* (Bartlett and Mendelson 2003b).

Theorem 3.2 *Let $\mathcal{D}$ be a distribution over $\mathcal{X} \times \mathcal{Y}$, satisfying*

$$\mathbb{P}_{(x,y)\sim\mathcal{D}} \left(|\{v : x^v \neq \bot\}| \neq 1 \right) = 0$$

Let $S = ((\mathbf{x}_i, y_i))_{i=1}^m$ be a dataset of m examples drawn i.i.d. according to $\mathcal{D}$. Let e be the 0/1 loss, and let $(\mathcal{F}_v)_{v=1}^V$ be the view-specific deterministic classifier sets. For each view v, denote $e \circ \mathcal{F}_v \overset{def}{=} \{(x^v, y) \mapsto e(f(x^v), y) | h \in \mathcal{F}_v\}$, and denote, for any sequence $S^v \in (\mathcal{X}_v \times \mathcal{Y})^{m_v}$ of size m_v, $R_{m_v}(e \circ \mathcal{F}_v, S^v)$ the empirical Rademacher complexity of $e \circ \mathcal{F}_v$ on S^v. Then, we have:

Baseline setting: for all $1 > \delta > 0$, with probability at least $1 - \delta$ over S:

$$\mathcal{E}(c^b_{f_1,...,f_V}) \leq \inf_{f'_V \in \mathcal{F}_v} \left[\mathcal{E}(c^b_{f'_1,...,f'_V}) \right] + 2 \sum_{v=1}^V \frac{m_v}{m} R_{m_v}(e \circ \mathcal{F}_v, S^v) + 6\sqrt{\frac{\ln(2/\delta)}{2m}}$$

where, for all v, $S^v \overset{def}{=} \{(x^v_i, y_i) | i = 1..m \text{ and } x^v_i \neq \bot\}$, $m_v = |S^v|$ and $f_v \in \mathcal{F}_v$ is the classifier minimizing the empirical risk on S^v.

multiview Gibbs classification setting: for all $1 > \delta > 0$, with probability at least $1 - \delta$ over S:

$$\mathcal{E}(c^{mg}_{f_1,...,f_V}) \leq \inf_{f'_V \in \mathcal{F}_v} \left[\mathcal{E}(c^b_{f'_1,...,f'_V}) \right] + \frac{2}{V} \sum_{v=1}^V R_m(e \circ \mathcal{F}_v, \underline{S}^v) + 6\sqrt{\frac{\ln(2/\delta)}{2m}} + \eta$$

where, for all v, $\underline{S}^v \overset{def}{=} \{(\underline{x}^v_i, y_i) | i = 1..m\}$, $f_v \in \mathcal{F}_v$ is the classifier minimizing the empirical risk on $\underline{S}^v$, and

$$\eta = \inf_{f'_V \in \mathcal{F}_v} \left[\mathcal{E}(c^{mg}_{f'_1,...,f'_V}) \right] - \inf_{f'_V \in \mathcal{F}_v} \left[\mathcal{E}(c^b_{f'_1,...,f'_V}) \right] \tag{3.50}$$

PROOF **of Theorem 3.2**
Fix a dataset S, and let $f_1, ..., f_V$ be the empirical risk minizers on each view, trained with the examples for which the corresponding view is observed. That is:

$$\forall v, f_v \in \underset{f \in \mathcal{F}_v}{\text{argmin}} \sum_{(x,y) \in S : x^v \neq \bot} e(f(x^v), y) \tag{3.51}$$

Considering that $c^b_{f_1,...,f_V}$ classifies an instance according to the classifier of the observed view, we can notice that $c^b_{f_1,...,f_V}$ is exactly the empirical risk minimizer over S for the set of classifiers $\mathcal{C}^b = \left\{ c^b_{f'_1,...,f'_V} | \forall v, f'_v \in \mathcal{F}_V \right\}$ (recall that exactly one view is observed for each example) with $f_v, v \in V$ defined as in Eq. (3.51), we have:

$$c^b_{f_1,...,f_V} \in \underset{c^b_{f'_1,...,f'_V} \in \mathcal{C}^b}{\text{argmin}} \sum_{(\mathbf{x},y) \in S} e\left(c^b_{f'_1,...,f'_V}(x^v), y\right)$$

Start then by using the well-known inequality that bounds the generalization error between the empirical risk minimizer and the error of the best-in class (see e.g. lemma 1.1 of Lugosi 2002):

$$\mathcal{E}\left(c^b_{f_1,...,f_V}\right) - \underset{c^b_{f'_1,...,f'_V} \in \mathcal{C}^b}{\inf} \left[\mathcal{E}\left(c^b_{f'_1,...,f'_V}\right)\right] \le 2$$

$$\underset{c^b_{f'_1,...,f'_V} \in \mathcal{C}^b}{\sup} \left| E\left(c^b_{f'_1,...,f'_V}, S\right) - \mathcal{E}\left(c^b_{f'_1,...,f'_V}\right)\right| \qquad (3.52)$$

where $E\left(c^b_{f'_1,...,f'_V}, S\right) = \frac{1}{|S|} \sum_{(\mathbf{x},y) \in S} e\left(c^b_{f'_1,...,f'_V}(\mathbf{x}), y\right)$ is the empirical risk, on the set S, of $c^b_{f'_1,...,f'_V}$.

Since e is the 0/1 loss, we have $0 \le e\left(c^b_{f'_1,...,f'_V}(\mathbf{x}), y\right) \le 1$ for any $(\mathbf{x}, y)$. We can thus use the standard Rademacher complexity analysis to obtain a data-dependent bound on the right-hand term of Eq 3.52, so we can apply McDiarmid's theorem McDiarmid (1989) to the function $S \mapsto \underset{c^b_{f'_1,...,f'_V} \in \mathcal{C}^b}{\sup} \left| E\left(c^b_{f'_1,...,f'_V}, S\right) - \mathcal{E}\left(c^b_{f'_1,...,f'_V}\right)\right|$, which can not change by more than $1/m$ when a pair $(\mathbf{x}, y)$ changes. We then have, with probability at least $1 - \delta/2$:

$$\underset{c^b_{f'_1,...,f'_V} \in \mathcal{C}^b}{\sup} \left| E\left(c^b_{f'_1,...,f'_V}, S\right) - \mathcal{E}\left(c^b_{f'_1,...,f'_V}\right)\right| \le$$

$$\mathbb{E}_{S' \sim \mathcal{D}^m} \underset{c^b_{f'_1,...,f'_V} \in \mathcal{C}^b}{\sup} \left| E\left(c^b_{f'_1,...,f'_V}, S'\right) - \mathcal{E}\left(c^b_{f'_1,...,f'_V}\right)\right| + \sqrt{\frac{\ln(2/\delta)}{2m}} \qquad (3.53)$$

where $m = |S|$.

We will now use the classical definition of the Rademacher complexity (see e.g. Bartlett and Mendelson 2003b) for some class of function $\mathcal{F}$ defined on the input space $\mathcal{X} \times \mathcal{Y}$:

$$\mathcal{R}_m(\mathcal{F}) \overset{def}{=} \mathbb{E}_{S' \sim \mathcal{D}^m} R_m(\mathcal{F}, S')$$

where $R_m(\mathcal{F}, S')$ is the empirical Rademacher complexity of $\mathcal{F}$ on the dataset S', defined as:

$$R_m(\mathcal{F}, S') \overset{def}{=} \mathbb{E}_{\sigma_1,...,\sigma_m} \underset{f \in \mathcal{F}}{\sup} \left| \frac{2}{m} \sum_{i=1}^{m} \sigma_i f((\mathbf{x}'_i, y'_i))\right|$$

where σ_i are independent random variables such that $\mathbb{P}(\sigma_i = 1) = \mathbb{P}(\sigma_i = -1) = 1/2$. With standard arguments of the Rademacher complexity analysis, we have, with probability at least $1 - \delta/2$:

$$\mathbb{E}_{S' \sim \mathcal{D}^m} \sup_{c^b_{f'_1,\ldots,f'_V} \in \mathcal{C}^b} \left| E\left(c^b_{f'_1,\ldots,f'_V}, S'\right) - \mathcal{E}\left(c^b_{f'_1,\ldots,f'_V}\right) \right|$$

$$\leq R_m(e \circ \mathcal{C}^b, S) + \sqrt{\frac{2 \ln(2/\delta)}{m}} \tag{3.54}$$

where $e \circ \mathcal{C}^b = \left\{ (\mathbf{x}, y) \mapsto e\left(c^b_{f'_1,\ldots,f'_V}(\mathbf{x}), y\right) \mid c^b_{f'_1,\ldots,f'_V} \in \mathcal{C}^b \right\}$. Plugging 3.54 into 3.53 and 3.52, we obtain, with probability at least $1 - \delta$:

$$\mathcal{E}\left(c^b_{f_1,\ldots,f_V}\right) \leq \inf_{c^b_{f'_1,\ldots,f'_V} \in \mathcal{C}^b} \left[\mathcal{E}\left(c^b_{f'_1,\ldots,f'_V}\right) \right] + 2R_m(e \circ \mathcal{C}^b, S) + 6\sqrt{\frac{\ln(2/\delta)}{2m}} \tag{3.55}$$

The final step consists in re-writing the empirical Rademacher complexity $R_m(e \circ \mathcal{C}^b, S)$ depending on the Rademacher complexity of the view-specific classifier sets. Given the dataset S, we define, for each view v, the partial dataset $S^v \overset{def}{=} \{(x^v_i, y_i) \mid i = 1..m \text{ and } x^v_i \neq \perp\}$. We use a specific index notation for the examples in $S^v = \{(x^v_{i^v_k}, y_{i^v_k}), k = 1..m_v\}$. By definition:

$$R_m(e \circ \mathcal{C}^b, S) = \mathbb{E}_{\sigma_1,\ldots,\sigma_m} \sup_{c^b_{f'_1,\ldots,f'_V} \in \mathcal{C}^b} \left| \frac{2}{m} \sum_{i=1}^m \sigma_i e\left(c^b_{f'_1,\ldots,f'_V}(\mathbf{x}_i), y_i\right) \right|$$

$$= \mathbb{E}_{\sigma_1,\ldots,\sigma_m} \sup_{c^b_{f'_1,\ldots,f'_V} \in \mathcal{C}^b} \left| \frac{2}{m} \sum_{v=1}^V \sum_{i:x^v_i \neq \perp} \sigma_i e\left(f'_v(x^v_i), y_i\right) \right|$$

$$\leq \sum_{v=1}^V \frac{m_v}{m} \mathbb{E}_{\sigma_1,\ldots,\sigma_{m_v}} \sup_{f'_v \in \mathcal{F}_v} \left| \frac{2}{m_v} \sum_{k=1}^{m_v} \sigma_k e\left(f'_v(x^v_{i^v_k}), y_{i^v_k}\right) \right|$$

$$= \sum_{v=1}^V \frac{m_v}{m} R_{m_v}(e \circ \mathcal{F}_v, S^v)$$

Together with Eq. 3.55 gives the desired result.

The proof for the multiview Gibbs classification setting follows the same steps, replacing $\mathcal{C}^b$ by $\mathcal{C}^{mg} \overset{def}{=} \left\{ c^{mg}_{f'_1,\ldots,f'_V} \mid \forall v, f'_v \in \mathcal{F}_v \right\}$. Only the last step (the calculation of the empirical Rademacher complexity) has to be modified. Remind that the true and empirical risks of the multiview Gibbs classifier are the average of empirical and true risks of the view-specific classifiers, as the multiview Gibbs classifier is supposed to be drawn from a uniform posterior distribution McAllester (2003). By the definition of the empirical Rademacher complexity, we have once again:

$$R_m(e \circ C^{mg}, S) = \mathbb{E}_{\sigma_1,\ldots,\sigma_m} \sup_{c^{mg}_{f'_1,\ldots,f'_V} \in C^b} \left| \frac{2}{m} \sum_{i=1}^{m} \sigma_i e \left(c^{mg}_{f'_1,\ldots,f'_V}(\mathbf{x}_i), y_i \right) \right|$$

$$= \mathbb{E}_{\sigma_1,\ldots,\sigma_m} \sup_{c^{mg}_{f'_1,\ldots,f'_V} \in C^b} \left| \frac{2}{mV} \sum_{v=1}^{V} \sum_{i=1}^{m} \sigma_i e \left(f'_v(\underline{x}^v_i), y_i \right) \right|$$

$$\leq \frac{1}{V} \sum_{v=1}^{V} \mathbb{E}_{\sigma_1,\ldots,\sigma_m} \sup_{f'_v \in \mathcal{F}_v} \left| \frac{2}{m} \sum_{i=1}^{m} \sigma_i e \left(f'_v(\underline{x}^v_i), y_i \right) \right|$$

$$= \frac{1}{V} \sum_{v=1}^{V} R_m(e \circ \mathcal{F}_v, \underline{S}^v)$$

where $\underline{S}^v \overset{def}{=} \{(\underline{x}^v_i, y_i) | i = 1..m\}$

This theorem gives a rule for whether it is preferable to learn only with the observed views (the baseline setting) or preferable to use the view-generating functions in the multiview Gibbs classification setting: we should use the former when $2 \sum_v \frac{m_v}{m} R_{m_v}(e \circ \mathcal{F}_v, S^v) < \frac{2}{V} \sum_v R_m(e \circ \mathcal{F}_v, \underline{S}^v) + \eta$, and the latter otherwise.

Consider first the role of η in Eq 3.50. The difference between the two settings is in the train and test distributions for the view-specific classifiers. η compares the best achievable error for each of the distribution. $\inf_{f'_v \in \mathcal{F}_v} \left[\mathcal{E} \left(c^b_{f'_1,\ldots,f'_V} \right) \right]$ is the best achievable error in the baseline setting (i.e. without generated views), with the automatically generated views, the best achievable error becomes $\inf_{f'_v \in \mathcal{F}_v} \left[\mathcal{E} \left(c^{mg}_{f'_1,\ldots,f'_V} \right) \right]$.

Therefore η measures the loss incurred by using the view generating functions. In a favorable situation, the quality of the generating functions will be sufficient to make η small.

The terms depending on the complexity of the class of functions may be better explained using orders of magnitude. Typically, the Rademacher complexity for a sample of size n is usually of order $O(\frac{1}{\sqrt{n}})$ (Bartlett and Mendelson 2003b).

Assuming, for simplicity, that all empirical Rademacher complexities in Theorem 3.2 are approximately equal to $d/\sqrt{n}$, where n is the size of the sample on which they are computed, and assuming that $m_v = m/V$ for all v. The trade-off becomes:

Choose the multiview Gibbs classification setting when: $d \left(\sqrt{\frac{V}{m}} - \frac{1}{\sqrt{m}} \right) > \eta$

This means that important performance gains is expected when the number of examples is small, the generated views of sufficiently high quality for the given classification task, and/or there are many views available.

Remark that one advantage of the multiview setting at prediction time is that it is possible to use a majority voting scheme. In such a case, it is expected that $\mathcal{E}\left(c^{mv}_{f'_1,...,f'_V}\right) \le \mathcal{E}\left(c^{mg}_{f'_1,...,f'_V}\right)$ if the view-specific classifiers are not correlated in their errors. It can not be guaranteed in general, though, since, in general, it is not possible to prove any better than $\mathcal{E}\left(c^{mv}_{f'_1,...,f'_V}\right) \le 2\mathcal{E}\left(c^{mg}_{f'_1,...,f'_V}\right)$ (see e.g. Langford and Shawe-taylor 2002).

3.4.2   Multiview Self-Training

One advantage of the multi-view settings described in the previous section is that unlabeled training examples may naturally be taken into account in a semi–supervised learning scheme, using existing approaches for multi-view learning (e.g. Blum and Mitchell 1998b).

In this section, we describe how, under the framework of Leskes (2005b), the supervised learning trade-off presented above can be improved using extra unlabeled examples. This framework is based on the notion of *disagreement* between the various view-specific classifiers, defined as the expected variance of their outputs:

$$\mathbb{V}\left(f_1, ..., f_V\right) \overset{def}{=} \mathbb{E}_{(\mathbf{x},y)\sim\mathcal{D}}\left[\frac{1}{V}\sum_v f_v(\underline{x}^v)^2 - \left(\frac{1}{V}\sum_v f_v(\underline{x}^v)\right)^2\right] \qquad (3.56)$$

The overall idea is that a set of good view-specific classifiers should agree on their predictions, making the expected variance small. This notion of disagreement has two key advantages. First, it does not depend on the true class labels, making its estimation easy over a large, unlabeled training set. The second advantage is that if, during training, it turns out that the view-specific classifiers have a disagreement of at most μ on the unlabeled set, the set of possible view-specific classifiers that needs be considered in the supervised learning stage is reduced to:

$$\mathcal{F}^*_v(\mu) \overset{def}{=} \left\{f'_v \in \mathcal{F}_v \,\middle|\, \forall v' \ne v, \exists f'_{v'} \in \mathcal{F}_{v'}, \mathbb{V}(f'_1, ..., f'_V) \le \mu\right\}$$

Thus, the more the various view-specific classifiers tend to agree, the smaller the possible set of functions will be. This suggests a simple way to do semi-supervised learning: the unlabeled data can be used to choose, among the classifiers minimizing the empirical risk on the labeled training set, those with best generalization performance (by choosing the classifiers with highest agreement on the unlabeled set). This is particularly interesting when the number of labeled examples is small, as the train error is usually close to 0.

Theorem 3 of Leskes (2005b) provides a theoretical value $B(\epsilon, \delta)$ for the minimum number of unlabeled examples required to estimate Eq. 3.56 with precision ϵ and probability $1 - \delta$ (this bound depends on $\{\mathcal{F}_v\}_{v=1..V}$). The following result gives a tighter bound of the generalization error of the multi-view Gibbs classifier when unlabeled data are available. The proof is similar to Theorem 4 in Leskes (2005b).

Proposition 3.2 *Let* $0 \le \mu \le 1$ *and* $0 < \delta < 1$. *Under the conditions and nota-tions of Theorem 3.2, assume furthermore that we have access to* $u \ge B(\mu/2, \delta/2)$ *unlabeled examples drawn i.i.d. according to the marginal distribution of* $\mathcal{D}$ *on* $\mathcal{X}$.

Then, with probability at least $1 - \delta$, *if the empirical risk minimizers* $f_v \in$ $\arg\min_{f \in \mathcal{F}_v} \sum_{(\underline{x}^v, y) \in \underline{S}^v} e(h, (\underline{x}^v, y))$ *have a disagreement less than* $\mu/2$ *on the unlabeled set, we have:*

$$\mathcal{E}\left(c^{mg}_{f_1,...,f_V}\right) \le \inf_{f'_v \in \mathcal{F}_v}\left[\mathcal{E}\left(c^{b}_{f'_1,...,f'_V}\right)\right] + \frac{2}{V}\sum_{v=1}^{V} R_m(e \circ \mathcal{F}^*_v(\mu), \underline{S}^v) + 6\sqrt{\frac{\ln(4/\delta)}{2m}} + \eta$$

We can now rewrite the trade-off between the baseline setting and the multi-view Gibbs classifier, taking semi-supervised learning into account. Using orders of magnitude, and assuming that for each view, $R_m(e \circ \mathcal{F}^*_v(\mu), \underline{S}^v)$ is $O(d_u/\sqrt{m})$, with the proportional factor $d_u \ll d$, the trade-off becomes:

Choose the mutli-view Gibbs classification setting when: $d\sqrt{V/m} - d_u/\sqrt{m} > \eta$.

Thus, the improvement is even more important than in the supervised setting. Also note that the more views we have, the greater the reduction in classifier set complexity should be.

Notice that this semi-supervised learning principle enforces agreement between the view specific classifiers. In the extreme case where they almost always give the same output, majority voting is then nearly equivalent to the Gibbs classifier (when all voters agree, any vote is equal to the majority vote). We therefore expect the majority vote and the Gibbs classifier to yield similar performance in the semi-supervised setting.

A simple extension of the self-training algorithm to the multiview case follow-ing this principle is provided in Algorithm 4. In this case, view-specific classifiers are first trained over the labeled training set and the thresholds for pseudo-labeling are then automatically computed according to their distributions of unsigned mar-gins estimated over the unlabeled training set. Each view-specific classifier assigns pseudo-class labels to unlabeled examples using the thresholds estimated at the pre-vious step and these pseudo-labels are finally decided according to a majority vote rule. New classifiers are then trained using the labeled training data and all the pseudo-labeled examples obtained from this two step pseudo-labeling procedure.

Algorithm 4: Multiview self-training algorithm

Input   : A learning algorithm: $\mathcal{A}$;

A labeled training set S and an unlabeled training set $X_{\mathcal{U}}$;

Initialization: $t \leftarrow 0$; $\tilde{S}_U \leftarrow \emptyset$; For each view v, train a Bayes classifier $B_{Q^v}^{(0)}$ on S using $\mathcal{A}$.

repeat

$\quad U \leftarrow \emptyset$;

$\quad$ Estimate the margin parameter $(\rho_v^{(t)})_{v=1}^V$;

$\quad$ **for** $\mathbf{x} = (x^1, ..., x^V) \in X_{\mathcal{U}}$ **do**

$$\forall v, \tilde{y}_v^{(t)} \leftarrow \begin{cases} \text{sgn}(B_{Q^v}^{(t)}) \text{if } B_{Q^v}^{(t)}(x^v) > \rho_v^{(t)} \\ 0 \text{ oterwise} \end{cases}$$

$\quad\quad$ **if** $\displaystyle\sum_{v=1}^V \tilde{y}_v^{(t)} > \frac{V}{2}$ **then**

$\quad\quad\quad \tilde{S}_U \leftarrow \tilde{S}_U \cup \{(\mathbf{x},+1)\}$;

$\quad\quad\quad U \leftarrow U \cup \{\mathbf{x}\}$;

$\quad\quad$ **else if** $\displaystyle\sum_{v=1}^V \tilde{y}_v^{(t)} < -\frac{V}{2}$ **then**

$\quad\quad\quad \tilde{S}_U \leftarrow \tilde{S}_U \cup \{(\mathbf{x},-1)\}$;

$\quad\quad\quad U \leftarrow U \cup \{\mathbf{x}\}$;

$\quad X_{\mathcal{U}} \leftarrow X_{\mathcal{U}} \setminus U$;

$\quad t \leftarrow t+1$;

$\quad$ For each view v, train a classifier $B_{Q^v}^{(t)}$ on $S \cup \tilde{S}_U$ with $\mathcal{A}$.

until $U = \emptyset \vee X_{\mathcal{U}} = \emptyset$;

Output   : View-specific Bayes classifiers $(B_{Q^v}^{(t)})_{v=1}^V$

In Sum

We have seen that

1. Semi-supervised approaches exploit the geometry of data to learn a prediction function,
2. Graphical approaches are based on an empirical graph constructed over the examples and reflecting their geometry,
3. Generative approaches could be inefficient in the case where the distributional hypothesis on which the generation of data is based are not met,
4. Discriminant approaches exploit the geometry of data using the unsigned margins of examples.

Chapter 4
Learning with Interdependent Data

Learning to rank, in which the goal is to order sets of objects unseen at training time, has emerged as major subfield of machine learning at the beginning of the 2000s (Cohen et al. 1998; Herbrich et al. 1998; Joachims 2002b). Early research on this topic was mainly motivated by the unprecedented scale of the problem of information overload caused by the recent explosion of the Web. Machine learning techniques appeared as good candidates to process, at the desired scale, the information collected about users and items, such as localization, browsing history or click-through data, in order to improve the relevance of results. Since information retrieval systems usually display their results in the form of a list of items ordered by predicted interest for the user, such machine learning techniques should be designed to effectively compare the relevance of different items.

Predicting a good ranking between objects is a different task than predicting an accurate assessment of their relevance. Even though one might argue that if the relevance of the objects were correctly predicted, then the induced ranking would also be correct, the difference lies in the more realistic scenario where predictions are not perfectly accurate. Figure 4.1 depicts an example where two different functions f_1 and f_2 are used to predict the relevance and to rank the same set of 2 relevant (squares, desired target value 1) and 3 irrelevant (ellipses, target value -1) examples. The first function f_1 is significantly better than f_2 in terms of squared error, but incorrectly ranks a negative example on the top, while f_2 predicts the desired ranking. Similar examples can also be found when measuring the classification error rather than the squared error. Such examples show that in order to evaluate and train ranking functions, criteria that measure the quality of the predictions on individual objects can be misleading. Learning to rank is more about comparing objects with each other on a relative scale than predicting the relevance of each individual object on an absolute scale.

Ranking tasks in machine learning are commonly dealt with in a pairwise comparison framework, where, given a desired order of the objects, the algorithm optimizes the proportion of pairs of objects for which the relative ordering is incorrect. Pairwise comparison approaches are convenient because they reduce the ranking problem to a binary classification problem over pairs of objects (x, x'), where a prediction of $+1$ means that x should be ranked higher then x'; the prediction is -1 otherwise. The

© Springer International Publishing Switzerland 2015
M.-R. Amini, N. Usunier, *Learning with Partially Labeled and Interdependent Data,*
DOI 10.1007/978-3-319-15726-9_4

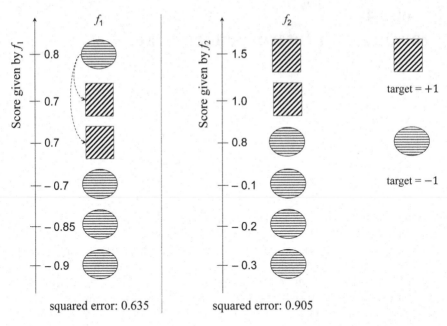

Fig. 4.1 Illustration of how better individual predictions on average can still be worse in terms of ranking. f_1 predicts the individual target values better than f_2 on average, but the ranking given by f_1 is incorrect

most popular algorithms for Ranking, such as RankBoost (Freund et al. 2003) or Ranking Support Vector Machines (Herbrich et al. 1998; Joachims 2002b) are built upon this reduction.

From a theoretical perspective however, learning to rank with pairwise comparisons presents at least two challenges:

1. Existing theoretical guarantees on the generalization error for classification are unapplicable to pairwise comparison approaches to ranking.
   The fundamental assumption for the statistical analysis of classification is the i.i.d. assumption. For ranking however, even if we assume that the objects to be ranked are i.i.d., a single object can appear in several pairs. The pairs of objects that are classified are thus not independent but rather *interdependent*.
2. The need to rank sets of objects can appear in many contexts that correspond to different theoretical framework.
   First, the need to predict a good ranking function might appear in problems that are naturally classification or regression tasks. For instance, if the target of a regression problem corresponds to some (monetary or non-monetary) value of the input instance, we might be interested in not only accurately predicting the value, but also be able to correctly order a set of test instances according to that value because we might be more interested in instances with higher value (or even interested only in them).

Secondly, the ranking problems that appear in search engines are of a different nature, because the task is to rank a set of candidate documents with respect to an input query. Each query defines its own ranking problem, and we need to generalize over multiple queries.

This chapter presents an extension of the framework of statistical learning, in the supervised setting, in order to analyze the generalization performance of classifiers that are trained and evaluated on examples that can be interdependent. The goal of this framework is two give an answer to the two problems above. First, the generalization error bounds we prove rely on a relaxed assumption that allows us to analyze criteria based on pairwise comparisons of objects. Second, our results do not make any prior restriction on the dependency structure between the objects to classify, but rather take the dependency structure as a parameter. This allows us to present different existing ranking frameworks in a unified way.

The main result of this chapter is an extension of the Rademacher complexity analysis (see Chap. 2) in a framework where the examples can be interdependent. We show how this analysis extends existing generalization error bounds for classification, and how some existing results on the capacity of a class of classifiers can be re-used when the examples are not independent. We also show the application of the general results to several ranking problems.

In the remainder of the chapter, we first present different frameworks under which ranking can be formally analyzed, and present how ranking can be reduced to a binary classification problem with interdependent data. We then present our unifying framework, and detail a concentration inequality of Janson (2004) for sums of partly dependent variables, which provides the fundamental tool underlying our analysis. The last section defines the fractional Rademacher complexity, an extension of the Rademacher complexity; we then prove a generic generalization error bound in our framework and show how it can be applied to two ranking problems that are very different in nature.

4.1 Pairwise Ranking Tasks

We describe in this section the two main frameworks of learning-to-rank: ranking of instances and ranking of alternatives. We describe the basic assumptions underlying the two frameworks, the prediction task and the performance criteria for training and evaluation that we study in this book, namely criteria based on comparisons of pairs of examples. We then make the link between ranking and the classification of pairs of objects, which is the motivation of the work presented in the next sections. We end the section with a brief description of works in learning to rank that do not fit in our framework.

The formal frameworks we consider in this chapter assume that the supervision takes the form of labels attached to each object to rank. Such type of feedback naturally appears naturally when the ranking problem is induced by a classification or regression problem, but also in practical applications such as search engines, where the relevance of a document to a query is assessed by annotators independently of other candidate documents, in both industry (Cossock and Zhang 2008; Chapelle and Chang 2011) and academia (Voorhees et al. 2005).

Remark 4.1 (Learning from pairwise preferences) We do not consider in this chapter the task of learning *from* pairwise preferences, which is commonly associated to ranking (Cohen et al. 1998; Freund et al. 2003). In such tasks, the training data comes in the form of a set of objects, together with a preference relation between objects. The preference relation given as supervision provides the relative desired ordering on a set of pairs of objects in the training set.

The framework of learning from pairwise preferences is usually less restrictive than our framework, but a practically useful theoretical result in that framework should consider realistic assumptions on how the preference information is collected. The reason why we do not consider pairwise preferences as feedback is that the motivation of the work presented here is the analysis of ranking problems, but in many cases of learning from pairwise preferences the theoretical problem to tackle is more related to this data collection process than to the problem of ranking.

For instance, a typical application of learning from pairwise preferences is to learn from implicit feedback, such as user clicks in search engines or recommender systems. In that case, pairwise preferences can be collected as follows: if a user traverses a list of recommended items or Web pages and if the user clicks on the k-th proposed item, then this clicked item is supposed to be preferred to the better ranked ones. No preference information is inferred from this click for items ranked after k, because the user might not have had the occasion to inspect them (Duchi et al. 2013). In that context, pairwise preferences are used as a correction to the bias created by the current recommendation system or search engine. However, the theoretical analysis of such biases is out of the scope of this chapter.

4.1.1   Ranking of Instances

The framework of ranking of instances was historically the object of the first studies in ranking (Cohen et al. 1998). Even though the framework was originally introduced for learning from pairwise preferences, a large portion of the subsequent works have focused on learning when individual labels (binary or real-valued) are given. We follow this assumption throughout this work.

When ranking instances, we assume that the train set $S = ((x_1, y_1), ..., (x_m, y_m))$ with $x_i \in \mathcal{X}$ and $y_i \in \mathbb{R}$ are sampled i.i.d. according to $\mathcal{D}$. As such, the basic assumption is the same as in classification or regression, but the goals are different. In ranking, the goal is to learn a scoring function $h : \mathcal{X} \to \mathbb{R}$ so that so that the ordering of new instances induced by h is in accordance with the ordering induced by their labels. That is, if (x, y) and (x', y') are two examples drawn from $\mathcal{D}$, we want $h(x) > h(x')$ when $y > y'$.

The most natural way to measure the true (generalization) ranking risk of a scoring function h is to measure the pairwise comparison error, and there are two common ways to measure it:

1. The probability that two examples (x, y) and (x', y') (each one drawn according to $\mathcal{D}$) are misordered by h (Clémençon et al. 2005):

$$\mathcal{E}^{pc}(f) = \mathbb{E}_{((x,y),(x',y'))\sim\mathcal{D}\times\mathcal{D}}\left[\mathbb{1}_{[y>y']}\mathbb{1}_{[f(x)\leq f(x')]}\right] \qquad (4.1)$$

Where pc in $\mathcal{E}^{pc}$ stands for pairwise comparisons. The empirical counterpart of this risk is then:

$$E^{pc}(f, S) = \frac{1}{m(m-1)} \sum_{i\neq j} \mathbb{1}_{[y_i>y_j]}\mathbb{1}_{[f(x_i)\leq f(x_j)]}. \qquad (4.2)$$

2. The conditional probability that two examples (x, y) and (x', y') are misordered, given that $y > y'$

$$\mathcal{E}^{cpc}(f) = \mathbb{E}_{((x,y),(x',y'))\sim\mathcal{D}\times\mathcal{D}}\left[\mathbb{1}_{[f(x)\leq f(x')]}|y > y\right]. \qquad (4.3)$$

so that the empirical risk associated to $\mathcal{E}^{cpc}$ is usually taken as:

$$E^{cpc}(f, S) = \frac{1}{\sum_{i\neq j}\mathbb{1}_{[y_i>y_j]}} \sum_{i\neq j} \mathbb{1}_{[y_i>y_j]}\mathbb{1}_{[f(x_i)\leq f(x_j)]}. \qquad (4.4)$$

The (true) ranking risks $\mathcal{E}^{pc}$ and $\mathcal{E}^{cpc}$ are equal up to the division by the probablity that $y > y$. Thus, the difference between the two is not important is theory, in the sense that when we compare two scoring functions, the one that is better for one risk is also better for the other risk. However, the empirical version E^{cpc} of $\mathcal{E}^{cpc}$ may be unbiased only asymptotically (see e.g. (Rudin et al. 2005; Rajaram and Agarwal 2005)).

While instance ranking in itself can be defined for any real-valued label (and more generally for pairwise preference feedback as in (Cohen et al. 1998; Freund et al. 2003; Clémençon et al. 2005; Rudin et al. 2005), among others), two important special cases have received particular attention. The first one is bipartite ranking, where the labels are either -1 or $+1$, and the second one is K-partite ranking where the labels are integers in the range $\{1, ..., K\}$.

Bipartite Ranking and the Area Under the ROC Curve

In bipartite ranking (Freund et al. 2003; Agarwal et al. 2005), the data is the same as in binary classification. The goal however is different, as in classification we search for correctly predicting the class label for each individual input, while in bipartite ranking the goal is to give higher scores to instances of the positive class than to instances of the negative class.

When evaluating with pairwise comparisons risks such as $\mathcal{E}^{pc}$ or $\mathcal{E}^{cpc}$, the Bayes-optimal predictors in bipartite ranking are all the scoring functions that are strictly increasing transforms of the posterior probability of class $+1$ (Clémençon et al. 2005).

In contrast, the Bayes-optimal binary classifier is to predict $+1$ if the posterior probability of class $+1$ is greater than 0.5. This remark helps us understand why bipartite ranking is strictly more difficult than binary classification: in binary classification, we only have to properly learn the level set 0.5 of the posterior probability function; however, in ranking, all the level sets of the posterior probability function have to be estimated (Clémençon and Vayatis 2010). From that point of view, ranking is more closely related to regression than to classification, and many regression-based approaches have been proposed for bipartite ranking (Kotlowski et al. 2011; Agarwal 2012).

Bipartite ranking has received a lot of attention because given a training set $S = ((x_1, y_1), ..., (x_m, y_m))$ with $y_i \in \{-1, 1\}$, the conditional pairwise comparison risk of f is equal to 1 minus the Area Under the ROC Curve (AUC) of f (Freund et al. 2003; Agarwal et al. 2005). Indeed, in that case, the empirical risk (4.4) can be written as:

$$E^{cpc}(f, S) = E^{AUC}(f, S) = \frac{1}{m^+ m^-} \sum_{i: y_i = 1} \sum_{j: y_j = -1} \mathbb{1}_{[f(x_i) \leq f(x_j)]}$$

$$\text{where } m^+ = \sum_{i=1}^{m} \mathbb{1}_{[y_i = 1]} \text{ and } m^- = \sum_{i=1}^{m} \mathbb{1}_{[y_i = -1]},$$

which is (1 minus) the Wilcoxon-Mann-Whitney statistics when f generates no ties, and is known to be equal to the AUC (Cortes and Mohri 2004). The true risk in bipartite ranking can then simply be written as the conditional probability that a positive example is ranked below a negative example:

$$\mathcal{E}^{AUC}(f) = \mathbb{E}_{((x,y),(x',y')) \sim \mathcal{D}^2}[\mathbb{1}_{[f(x) \leq f(x')]} | y = 1, y' = -1]. \tag{4.5}$$

Even though the AUC is intrinsically a ranking criterion, it is a usual criterion for the performance of classifiers when the classes are imbalanced. Cortes and Mohri (2004) have shown that directly optimizing the AUC instead of classification accuracy can lead to substantially better performance in terms of AUC when the classes are imbalanced. These early positive results on ranking have been one of the major motivation for optimizing ranking criteria even outside the scope of pure ranking problems, in the sense that ranking of instances can also be a means to build better classifiers.

K-Partite Ranking

A second important special case is K-partite ranking, where labels correspond to discrete ordinal levels, identified with $1, ..., K$ here. The data is then the same as in ordinal regression (Crammer and Singer 2002; Rajaram and Agarwal 2005; Clémençon et al. 2013), but once again the goal is only to order instances according to their ordinal labels rather than predicting the label for each instance.

In general, assessing the performance in ordinal regression is difficult because, by definition, the cost of predicting y' instead of y is unknown; the only specification of the task is that the cost of predicting $y - 2$ instead of y is greater than the cost of $y - 1$ (and the same holds with $y + 2$ and $y + 1$ instead of y). The problem of assessing the quality of a scoring function in K-partite ranking suffers from the same problem, and several criteria have been devised. The first cases are based on multi-dimensional extensions of the ROC curve (Hand and Till 2001), such as the volume under the ROC surface of Clémençon et al. (2013). Another, more usual, criterion is the pairwise comparison errors $\mathcal{E}^{cpc}$ or $\mathcal{E}^{pc}$, with additional costs $a_{k,k'}$ (e.g. $a_{k,k'} = k' - k$) that specify the cost of ranking an instance of label k' above an instance of label k when $k' > k$ (Rajaram and Agarwal 2005).

$$\mathcal{E}^{cpc}_{k-part}(f) = \sum_{k'>k} \rho_k \rho_{k'} \mathbb{E}_{((x,y),(x',y'))\sim\mathcal{D}^2} [a_{k,k'} \mathbb{1}_{[f(x')\leq f(x)]} | y' = k', y = k],$$

where ρ_k is the marginal probability of label k.

The empirical risk associated to this cost-based pairwise error is then:

$$E^{cpc}_{k-part}(f, S) = \frac{1}{\sum_{i\neq j} \mathbb{1}_{[y_i>y_j]}} \sum_{i\neq j} \mathbb{1}_{[y_i>y_j]} a_{y_j,y_i} \mathbb{1}_{[f(x_i)\leq f(x_j)]}.$$

Even though E^{cpc}_{k-part} is a biased estimate of $\mathcal{E}^{cpc}_{k-part}$, it has been shown that the estimate is asymptotically unbiased (Rajaram and Agarwal 2005).

4.1.2 Ranking of Alternatives

In the previous subsection, we presented the task of ranking instances. Another major topic in learning to rank is what we call here ranking of alternatives, which encompasses the problems of label ranking (Dekel et al. 2003) or subset ranking (Cossock and Zhang 2008). In that setting, each *instance* is structured, in the sense that it contains a set of items, and defines a ranking problem in itself.

This kind of structure in the data corresponds to the "query-item" structure that can be found in learning to rank for Web information retrieval, which can be described as follows. Each user query is an *instance* of the learning problem; the search engine operates a first preselection of candidate Web pages (the alternatives), and the scoring function should sort these Web pages according to their predicted relevance to the user query. Thus, in this type of ranking, we need generalize to new queries, that is to new ranking problems. A scoring function should now give a score to alternatives depending on the query, and the risk of the scoring function is the average of the ranking risks incurred over different queries.

In this information retrieval scenario, the score is usually predicted based on joint features of the query and the alternatives. These joint features can be similarity between the content or title of the document and the keywords of the query (e.g. number of words in common), query-document matching ranking heuristics from Information Retrieval like BM25 or tf.idf (Manning et al. 2008). The goal is then to learn a function that gives a score to each document based on these features.

In the general context of ranking of alternatives, the unit of data that is sampled i.i.d. is a query, the set of alternatives (e.g. candidate documents) for that query, and, in the supervised learning setting, the relevance judgments that have been collected for each document. This can be modeled with two sequences $(\mathbf{x} = (x^1, ..., x^{n(\mathbf{x})}), \mathbf{y} = (y^1, ..., y^{n(\mathbf{x})}))$, where $n(\mathbf{x})$ is the number of alternatives for the query, $(x^1, ..., x^{n(x)}) \in \mathcal{X}$ are the joint feature vectors between the query and the alternatives (one feature vector for each alternative). Then, the loss of the scoring function $h : \mathcal{X} \to \mathbb{R}$ is any of the pairwise comparison errors described in the last subsection, applied to $((x^1, y^1), ..., (x^{n(\mathbf{x})}, y^{n(\mathbf{x})}))$. For instance with the conditional pairwise comparison loss, the generalization error can be written as:

$$\mathcal{E}_{alt}^{cpc}(f) = \mathbb{E}_{(\mathbf{x},\mathbf{y}) \sim \mathcal{D}} \left[\frac{1}{\sum_{j \neq k} \mathbb{1}_{[y^j > y^k]}} \sum_{j \neq k} \mathbb{1}_{[y^j > y^k]} \mathbb{1}_{[f(x^j) \leq f(x^k)]} \right],$$

and the empirical error on a training set $S = ((\mathbf{x}_1, \mathbf{y}_1), ..., (\mathbf{x}_m, \mathbf{y}_m))$ is defined by:

$$E_{alt}^{cpc}(f, S) = \frac{1}{m} \sum_{i=1}^{m} \frac{1}{\sum_{j \neq k} \mathbb{1}_{\left[y_i^j > y_i^k\right]}} \sum_{j \neq k} \mathbb{1}_{\left[y_i^j > y_i^k\right]} \mathbb{1}_{\left[f(x_i^j) \leq f(x_i^k)\right]}.$$

Multiclass Classification and Label Ranking

The framework for ranking of alternatives is not restricted to information retrieval scenarios, but also to data that are classified in one or several classes when there are more than two classes (multiclass and multilabel classification). Indeed, for any multiclass problem, we can define a *label ranking* task where the set of possible class labels has to be ranked for each input x (Dekel et al. 2003). The goal is then to predict on the top of the list the class(es) to which x belongs.

Compared to the framework defined above, we could view each input x of a multiclass or multilabel problem as a vector $\mathbf{x} = (x^1, ..., x^K)$ where K is the total number of labels, and x^k would be a joint feature representation of the original input x and class label k. In practice however, one does not have any such joint features between an input and a class label, and one rather learns one scoring function per class label. Nonetheless, we can still cast such problems in the framework of ranking of alternatives described above, by defining $\mathbf{x}$ and the x^k as a function of the original input as follows:

- assume the original input $x \in \mathbb{R}^d$ and consider it as a row vector,
- for each class label $k \in \{1, ..., K\}$, define x^k as a $K \times d$-dimensional vector containing 0 everywhere but in dimensions $\{(k-1)d+1, ..., kd\}$ where it is equal to x:

$$x^1 = [x \; 0_d \; 0_d \; ... \; 0_d],$$
$$x^2 = [0_d \; x \; 0_d \; ... \; 0_d],$$

$$\cdots$$

$$x^K = [0_d \quad \cdots \quad 0_d \ x],$$

where 0_d is the row vector of $\mathbb{R}^d$ containing only 0s.

The advantage of this construction is that learning a unique scoring function $f :$ $\mathbb{R}^{K \times d} \to \mathbb{R}$ and ranking class labels $\{1, ..., K\}$ according to $f(x^1), ..., f(x^K)$ is equivalent to learning K different functions $f^1, ..., f^K : \mathbb{R}^d \to \mathbb{R}$ and ranking the labels $\{1, ..., K\}$ according to $f^1(x), ..., f^K(x)$. Consequently, the framework of ranking of alternatives described in this section can capture label ranking tasks as well as query/document ranking tasks.

Remark 4.2 (Formula for the conditional pairwise comparison risk) Throughout this chapter, we use label ranking in multiclass classification (i.e. when an example (x, y) has the property that the input x belongs to one and only one of the classes $\{1, ..., K\}$), and in that respect we need specific notation. First, we consider the label ranking in the multiclass case as a problem of ranking of alternatives where each example $(\mathbf{x}, \mathbf{y})$ is constructed from the original data as follows:

1. $\mathbf{x} = (x^1, ..., x^K) \in (\mathbb{R}^{Kd})^K$ is defined as we previously described,
2. $\mathbf{y} = (y^1, ..., y^K) \in \{0, 1\}^K$ is a vector containing exactly one 1 at position k, where k is the desired class label of the input.

Given $\mathbf{y} \in \{0, 1\}^K$ with a single non-zero value, it will be convenient in Sect. 4.2.1 to reindex the class labels as a function of the desired labels. To that end, we use the function μ is defined as follows:

$$\forall k \in \{1, ..., K\} : \mu(\mathbf{y}, k) = \begin{cases} k' \text{such that } y^{k'} = 1 & \text{if } k = 1 \\ k & \text{if } k \neq \mu(\mathbf{y}, 1) \\ 1 & \text{if } k = \mu(\mathbf{y}, 1) \end{cases}.$$

Stated otherwise, the reindexing function swaps class label 1 and the desired class label. The reindexing function allows us to rewrite the empirical conditional pairwise risk for multiclass classification:

$$E^{mc}(f, S) = \frac{1}{m(K - 1)} \sum_{i=1}^{m} \sum_{k=2}^{K} \mathbb{1}_{\left[f(x_i^{\mu(y_i, 1)}) \leq f(x_i^{\mu(y_i, k)}) \right]}. \tag{4.6}$$

This formula of the conditional pairwise comparison risk for multiclass classification will be used when describing the framework of classification with interdependent data in Sect. 4.2.1.

4.1.3   Ranking as Classification of Pairs

The risk functions defined above, in both frameworks of ranking of instances or ranking of alternatives, are based on the comparisons of pairs of objects (either pairs of instances or pairs of alternatives), and in that sense we could interpret the training criteria above as learning a binary classifier of pairs of objects (x, x'), for which the binary decision is whether x should be ranked higher or lower than x'. The goal of this subsection is to formalize the similarities and differences between ranking and classification. In particular, we will see that binary classification algorithms, especially when the scoring function is a linear combination of the features, can immediately be transformed into learning-to-rank algorithms. On the other hand, we show that the binary classification theory cannot be readily applied to learning to rank, because the i.i.d. assumption of the examples to be classified is violated. In the remainder of this subsection, we only consider the case of instance ranking, and discuss the ranking of alternatives at the end.

Let us consider the case of instance ranking. Let $f : \mathcal{X} \to \mathbb{R}$ be a scoring function. Then, the classifiers of pairs associated to f, which is denoted by $\tilde{f}$, is defined by:

$$\forall x, x' \in \mathcal{X}, \tilde{f}(x, x') = 2\mathbb{1}_{[f(x) - f(x') > 0]} - 1,$$

which means that $\tilde{f}(x, x')$ is 1 when f ranks x higher than x', and -1 in the other case.

Now, if we consider the definition of the conditional pairwise comparison error of (4.3), we can see that:

$$\mathcal{E}^{cpc}(f) = \mathbb{E}_{((x,y),(x',y')) \sim \mathcal{D} \times \mathcal{D}}[\mathbb{1}_{[f(x) \leq f(x')]} | y > y]$$
$$= \mathbb{E}_{((x,y),(x',y')) \sim \mathcal{D} \times \mathcal{D}}[\mathbb{1}_{[\tilde{f} \neq 1]} | y > y'].$$

Then, $\mathcal{E}^{cpc}(f)$ is exactly the generalization error, in terms of *classification error* of $\tilde{f}$ on pairs of examples (x, x'), where $((x, y), (x', y'))$ follow the distribution $\mathcal{D} \times \mathcal{D}$ conditioned on $y > y'$.

From an algorithmic perspective, this relationship between ranking and pairwise classification is at the root of many algorithms. We detail below the cases of linear scoring functions first, and then briefly discuss the case of non-linear functions.

Algorithms for Learning Linear Scoring Functions

When the instances x lie in a (finite dimensional) vector space, say $\mathbb{R}^d$, we can learn scoring functions of the form $f(x) = \langle w, x \rangle$ where $\langle ., . \rangle$ is the canonical dot product and w is the parameter vector to be learnt. Then, using the linearity of $\langle w, . \rangle$, the classifier of pairs $\tilde{f}$ can be written as:

$$\forall x, x' \in \mathcal{X} = \mathbb{R}^d, \tilde{f}(x, x') = 2\mathbb{1}_{[f(x) - f(x') > 0]} - 1 = 2\mathbb{1}_{[\langle w, x - x' \rangle > 0]} - 1. \quad (4.7)$$

Thus, in the linear case, the classification (and thus ranking) of a pair $(x, x') \in \mathbb{R}^d$ is the binary classification of a unique vector of $\mathbb{R}^d$ defined by $x - x'$. Consequently, we can learn a linear scoring function on a training set $S = ((x_1, y_1), ..., (x_m, y_m))$ by applying *any* learning algorithm for binary classification with linear functions to the transformed training set $\mathfrak{T}^{pairs}(S) = ((x_i - x_j, 1)_{i,j:y_i>y_j})$, where the 1 corresponds to the binary label of each pair, which is constant by construction. Those pairs $(x, y), (x', y')$ that are included in $\mathfrak{T}^{pairs}(S)$ (i.e. those with $y > y'$) are called the *crucial pairs*. This approach of applying a binary classification algorithm to $\mathfrak{T}^{pairs}(S)$ is used in Support vector machines (SVMs) for Ranking (Herbrich et al. 1998; Joachims 2002b), and efficient implementations that take into account the pairwise structure of the training set have been developed (e.g. Joachims (2005)). More generally, logistic regression or ridge regression can be applied to $\mathfrak{T}^{pairs}(S)$ in practice to learn the parameter vector of a linear scoring function.

Learning Non-linear Scoring Functions

Algorithms for learning non-linear scoring functions can also be easily derived from existing algorithms. For instance kernel methods for support vector machines can be extended to pairs: if κ is a kernel over the input space $\mathcal{X}$, we can build the kernel $\tilde{\kappa}$ over the product space $\mathcal{X} \times \mathcal{X}$:

$$\forall((x, x'), (z, z')) \in (\mathcal{X} \times \mathcal{X})^2,$$
$$\tilde{\kappa}((x, x'), (z, z')) = \kappa(x, z) - \kappa(x', z) - \kappa(x, z') + \kappa(x', z').$$

That way, it can be easily checked that $\tilde{\kappa}$ represents a pair (x, x') of instances by the difference of the projections of x and x' in the reproducing kernel Hilbert space of κ.

Kernel methods for ranking suffer from a high computational complexity though, because the number of pairs in $\mathfrak{T}^{pairs}(S)$ is still quadratic in the number of samples in the original training set S. Since kernel methods usually have a running time that is quadratic with respect to the number of support vectors, this leads to an unacceptable running time of order m^4. Consequently, kernel methods have received little attention from the learning to rank community, and more focus has been given to algorithms with more efficient prediction/learning time. Scoring functions based on ensemble of trees trained with gradient boosting of regression trees (Zheng et al. 2008; Chapelle and Chang 2011), directly optimizing a squared error between $f(x) - f(x')$ and $y - y$, or more simply between $f(x)$ and y, have received a lot of acceptance by the community. These methods, in particular, have also exhibited state-of-the-art results for the ranking errors that are not based on pairwise comparisons that we present in Sect. 4.1.4.

Statistical Analysis of Pairwise Comparison Errors

While the framework of pairwise comparisons allows to derive algorithms for training scoring functions using existing classification algorithms, the theoretical analysis of pairwise ranking criteria is less straightforward. Indeed, the generalization error bounds for classification presented in Chap. 2 rely on the assumption that the inputs of the classifier are i.i.d. samples from a fixed distribution. However, the set of crucial pairs $\mathfrak{T}^{pairs}(S)$ built upon the original training set S does not contain independent samples, even if the examples of S are i.i.d. samples from a fixed distribution. Indeed, a single example $(x, y) \in S$ will appear in all the pairs $\{(x, x')|(x', y') \in S : y' > y\}$ of $\mathfrak{T}^{pairs}(S)$, creating a dependency between the different pairs of $\mathfrak{T}^{pairs}(S)$. As such, the generalization analysis for the binary classification of i.i.d. instances cannot be readily applied to ranking; the goal of the rest of this chapter is to extend the existing statistical analysis of classification with i.i.d. examples to ranking (i.e. classification of pairs), and to show that capacity measures used in binary classification are still relevant to ranking.

4.1.4 Other Ranking Frameworks

In this book, we consider only ranking error measures based on pairwise comparisons. Such a criterion based on the classification of pairs of examples corresponds to learning a (pre-)order over the instance space, and thus corresponds to the basic formulation of ranking. In addition, the reduction to the classification of pairs suggest to learn scoring functions with algorithms based on pairwise comparisons.

Nonetheless, other approaches exist for learning scoring functions. For both frameworks of ranking of instances or alternatives, different training schemes that do not consider pairs of examples are advocated. Also, different definitions of risk have been investigated than those based on pairwise comparisons. We briefly describe here some of these approaches, starting with those designed in the instance ranking framework.

In instance ranking, bipartite ranking has been extensively studied. In particular, in terms of learning algorithms, a number of recent works have advocated for using standard logistic or squared loss regression in the case of the AUC risk in bipartite ranking (Agarwal 2012; Kotlowski et al. 2011). The arguments are that such regression approaches are asymptotically optimal, and may be competitive with usual (pairwise comparison-based) learning-to-rank algorithms in practice. Their practical advantage is computational complexity, since they do not require to implicitly or explicitly construct pairs of examples. Such approaches are called *pointwise* (Liu 2009) because they consider instances individually.

In terms of ranking error criteria for bipartite ranking, the Area Under the ROC curve (in our framework, the conditional pairwise comparison risk) is usually the criterion of interest. However, several other performance measures have been studied, in order to put greater emphasis on the top ranked instances during learning and evaluation. In that context, the precision/recall (PR) curves and approximations of the area under PR curves traditionally used in information retrieval (Manning et al. 2008) might be considered; however, there is no learning algorithm to optimize them.

Several authors have investigated other definitions of ranking risk than the AUC. These risks have the same goal of obtaining a better ranking on the top as precision/recall analysis, but are more amenable to practical optimization and theoretical analysis. For instance, Rudin (2009) studied a criterion called the p-norm push, in order to penalize negative examples that are ranked on the top of the list. The empirical risk associated to this p-norm push can be defined as follows:

$$E^{p-norm}(f, S) = \left(\frac{1}{m^-} \sum_{j:y_j=-1} \left(\frac{1}{m^+} \sum_{i:y_i=1} \mathbb{1}_{[f(x_i) \leq f(x_j)]} \right)^p \right)^{1/p}$$

Note that the p-norm push is only meaningful in the bipartite case, and is thus an alternative to the AUC risk (4.5). Other criteria involve the functional approximation of the ROC curve of Clémençon and Vayatis (2009), or linear rank statistics (Clémençon and Vayatis 2007).

In the framework of ranking of alternatives, the attention has mostly been given to optimizing performance measures of search engines. Among those, the Discounted Cumulative Gain (DCG) (Järvelin and Kekäläinen 2002), which is conceptually similar to the linear rank statistics (Clémençon and Vayatis 2007) in bipartite ranking. The idea of the DCG is to define a gain for each example (x, y) as a function of the label y, and to sum the gains by traversing the ranked list of objects, with a discount factor that depends on the position in the list.

In order to give the formal definition of the DCG, we remind the reader that an example in that framework is a couple $(\mathbf{x}, \mathbf{y})$ where $\mathbf{x} = (x_1, ..., x_{n(\mathbf{x})})$ and $\mathbf{y} = (y_1, ..., y_{n(\mathbf{x})})$. In the equation below, we denote $\tilde{f}_k(x)$ the index in $1, ..., n(\mathbf{x})$ of instance ordered at rank k by f. We also denote γ_k is the discount applied to the gain rank k, while $G(()y)$ is the gain associated to label y. Usually, $\gamma_k = \frac{1}{\log(k+1)}$ and $G(y) = 2^y - 1$ in information retrieval (Järvelin and Kekäläinen 2002; Manning et al. 2008). Notice that this performance measure should be *maximized* (i.e. higher is better, in contrast to pairwise comparison errors that should be minimized), and that in practice the gain is summed only until a fixed rank K, ignoring the ranking after that rank cutoff K. In Search engines where the user in only shown the ten top results, the usual value is $K = 10$.

$$\mathcal{E}^{DCG}(f) = \mathbb{E}_{(\mathbf{x}, \mathbf{y}) \sim \mathcal{D}} \left[\sum_{k=1}^{\min(n(\mathbf{x}), K)} \gamma_k G(y_{\tilde{f}_k(\mathbf{x})}) \right].$$

Learning algorithms based on minimizing pairwise comparison errors are often considered as suboptimal for such error measures Liu (2009), and so-called listwise algorithms, which consider interactions of scores on the entire ranked list rather than just by pairs, are advocated by some authors (Quoc and Le 2007; Burges 2010; Cao et al. 2007). Nonetheless, algorithms based on ensemble methods and plain squared loss regression or variants of them have shown to have good theoretical properties and good empirical performances as well for these criteria (Cossock and Zhang 2008; Ravikumar et al. 2011; Calauzènes et al. 2013; Chapelle and Chang 2011). In this book, we do not consider the statistical analysis of such measures. We rather focus on the analysis on criteria based on pairwise comparisons and we analyze the relationship between such criteria and binary classification.

4.2   Classification of Interdependent Data

The core of our work is to analyze the generalization error of classification algorithms that are trained on non-i.i.d. samples. This analysis is motivated by the pairwise-comparison-based ranking problems described previously, but can be applied to derive error bounds for other problems such as multiclass classification that can also be reduced to the classification of pairs (Har-Peled et al. 2002).

The i.i.d. assumption is critical in statistical learning theory because of the use of concentration inequalities such as Hoeffding bound or McDiarmid's inequality. Consequently, dealing with non-i.i.d. examples require other concentration results than these inequality. Our work is based on a concentration inequality for sums of partly dependent random variables by Janson (2004). This result is particularly relevant to our analysis because it leverages the knowledge of the dependency structure of random variables, and this dependency structure between pairs of examples is indeed known in advance in learning to rank.

In this section, we describe first a formal framework to analyze classification problems with interdependent data. Then, we present Janson's concentration inequality (Janson 2004), which is the basis of our work. At the end of the section, we give the examples of how training set of pairs for bipartite and multiclass classification can be cast in the dependency structure defined by Janson (2004).

4.2.1   Formal Framework of Classification with Interdependent Data

The formal framework we consider for learning with interdependent data is the following: given a training set of examples $S = ((x_1, y_1), ..., (x_n, y_n))$ where each $(x_i, y_i) \in \mathcal{X} \times \mathcal{Y}$ of *independent* but non necessarily identically distributed samples, there is a fixed transformation function $\mathfrak{T}$:

$$\mathfrak{T} : (\mathcal{X} \times \mathcal{Y})^m \to (\{0, 1\} \times \tilde{\mathcal{X}} \times \{-1, 1\})^M \,,$$

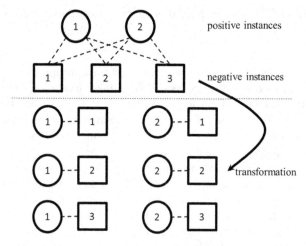

Fig. 4.2 Representation of the transformation of the examples in the case of bipartite ranking

that creates a new training set $\mathfrak{T}(S)$ of size M. This new training set contains M examples $((\tilde{z}_i, \tilde{x}_i, \tilde{y}_i))_{i=1}^M$, where:

- $\tilde{z}_i$ is either 0 or 1, and gives the information of whether $(\tilde{z}_i, \tilde{x}_i, \tilde{y}_i)$ should be included or ignored in the computation of the error,
- $\tilde{x}_i \in \tilde{\mathcal{X}}$ is the representation of the input for the new classification problem,
- $\tilde{y}_i \in \{-1, 1\}$ is the desired (binary) label for $\tilde{x}_i$.

The goal is to learn a binary classifier for the transformed examples $\tilde{f} : \tilde{\mathcal{X}} \to \{-1, 1\}$. As an immediate example, when ranking instances, $\tilde{x}_i$ is one pair of the pairs created from the original examples, so that $\tilde{\mathcal{X}} = \mathcal{X}^2$. The weights $\tilde{z}_i$ allow us to consider non-conditional pairwise risks such as (4.2) in instance ranking, where a pair (x, x') built upon two examples (x, y) and (x', y') is considered only when $y > y'$.

Concrete examples of transformation functions are given below, but we first give the formal expression of the true and empirical error in that framework. At first, we do not assume that the original examples are identically distributed (they must be independent samples, however). That is a technical detail that can be ignored for now; we shall see that it allows us to capture cases where the number of transformed examples M depends on the actual value of the original labels as in bipartite ranking.

To describe the formal framework in its most general formulation with independent but non-identically distributed original examples, we denote by $\mathcal{D}_i$ the distribution of the i-th original example (x_i, y_i), and by $\mathcal{D}_1^m$ the product measure $\mathcal{D}_1 \otimes \mathcal{D}_2 \otimes \ldots \otimes \mathcal{D}_m$, which is the distribution of S. Then, the empirical risk with interdependent data is then written as:

$$E^{\mathfrak{T}}(\tilde{f}, S) = \frac{1}{M} \sum_{i=1}^M \tilde{z}_i \, \mathbb{1}_{\left[\tilde{f}(\tilde{x}_i) \neq \tilde{y}_i\right]}$$

where $((\tilde{z}_1, \tilde{x}_1, \tilde{y}_1), \ldots, (\tilde{z}_M, \tilde{x}_M, \tilde{y}_M)) = \mathfrak{T}(S),$

and the true risk of $\tilde{f}$ is simply defined as the expected value of the empirical risk:

$$\mathcal{E}^{\mathfrak{T}}(\tilde{f}) = \mathbb{E}_{S \sim \mathcal{D}_1^m}[E^{\mathfrak{T}}(\tilde{f}, S)]. \tag{4.8}$$

In our applications, the transformation function is the mapping between the original examples and the set of pairs created from these examples. For example, in the framework of ranking of instances, we then have $\tilde{\mathcal{X}} = \mathcal{X}^2$. Figure 4.2 gives a pictorial representation of the pairs created by the transformation in the case of bipartite ranking. In the figure, there are two positive instances (circles) and 3 negative instances (squares). The complete bipartite graph on the top shows all pairs that have to be created. Then, the transformation generates these pairs where each positive example is compared to a negative example. The classifier we learn is the classifier of pairs induced by the scoring function. We note at this point that we did not assume each of the examples to be identically distributed. This is a technical detail for now, yet it is important when the number and/or set of pairs to consider depends on the observed values of the labels. We give below examples of transformation for the pairwise comparison risk in ranking of instances, the AUC risk in bipartite ranking, and the conditional pairwise comparison risk in ranking of alternatives in the special case of multiclass classification.

EXAMPLE **Pairwise Comparison Risk (4.1) in Ranking of Instances**
Considering the case of the non-conditional ranking risk based on pairwise comparisons (4.1), the empirical risk associated to a scoring function f is given by:

$$E^{pc}(f, S) = \frac{1}{m(m-1)} \sum_{i \neq j} \mathbb{1}_{[y_i > y_j]} \mathbb{1}_{[f(x_i) \leq f(x_j)]},$$

and the true risk is its expected value over i.i.d. sampled training sets.
Now, if we take $M = m(m-1)$ and define $\mathfrak{T}(S)$ to be equal to:

$$(\underbrace{\mathbb{1}_{[y_1 > y_2]}}_{\tilde{z}_1}, \underbrace{(x_1, x_2)}_{\tilde{x}_1}, \underbrace{1}_{\tilde{y}_1}), (\underbrace{\mathbb{1}_{[y_1 > y_3]}}_{\tilde{z}_2}, \underbrace{(x_1, x_3)}_{\tilde{x}_2}, \underbrace{1}_{\tilde{y}_2}) ..., (\underbrace{\mathbb{1}_{[y_1 > y_m]}}_{\tilde{z}_{m-1}}, \underbrace{(x_1, x_m)}_{\tilde{x}_{m-1}}, \underbrace{1}_{\tilde{y}_{m-1}}),$$

$$(\underbrace{\mathbb{1}_{[y_2 > y_1]}}_{\tilde{z}_m}, \underbrace{(x_2, x_1)}_{\tilde{x}_m}, \underbrace{1}_{\tilde{y}_m}), (\underbrace{\mathbb{1}_{[y_2 > y_3]}}_{\tilde{z}_{m+1}}, \underbrace{(x_2, x_3)}_{\tilde{x}_{m+1}}, \underbrace{1}_{\tilde{y}_{m+1}}) ..., (\underbrace{\mathbb{1}_{[y_2 > y_m]}}_{\tilde{z}_{2m-1}}, \underbrace{(x_2, x_m)}_{\tilde{x}_{2m-1}}, \underbrace{1}_{\tilde{y}_{2m-1}}),$$

$$...$$

$$(\underbrace{\mathbb{1}_{[y_m > y_1]}}_{\tilde{z}_{(m-1)^2+1}}, \underbrace{(x_m, x_1)}_{\tilde{x}_{(m-1)^2+1}}, \underbrace{1}_{\tilde{y}_{(m-1)^2+1}}) ..., (\underbrace{\mathbb{1}_{[y_m > y_{m-1}]}}_{\tilde{z}_{m(m-1)}}, \underbrace{(x_m, x_{m-1})}_{\tilde{x}_{m(m-1)}}, \underbrace{1}_{\tilde{y}_{m(m-1)}})),$$

then, taking $\tilde{f}$ as the classifier of pairs associated to the scoring function f (defined in (4.7)) such that $\tilde{f}(x, x') \neq 1 \Leftrightarrow \mathbb{1}_{[f(x) \leq f(x')]}$ we can see that:

$$E^{\mathfrak{T}}(\tilde{f}, S) = \frac{1}{M} \sum_{i=1}^{M} \tilde{z}_i \, \mathbb{1}_{[\tilde{f}(\tilde{x}_i) \neq \tilde{y}_i]}$$

$$= \frac{1}{m(m-1)} \sum_{i \neq j} \mathbb{1}_{[y_i > y_j]} \mathbb{1}_{[f(x_i) \leq f(x_j)]} = E^{pc}(f, S).$$

EXAMPLE **Area Under the ROC Curve**

The AUC risk, or, equivalently, the conditional pairwise comparison risk in bipartite ranking (4.5), can also be written as a classification of interdependent pairs in our framework.

First, following Agarwal et al. (2005), we notice that concentration inequalities, and generalization error bounds for the AUC risk can be derived by conditioning the training set to class labels as follows: let $\mathcal{D}_{+1}$ be the conditional distribution over the input space of positive examples, and $\mathcal{D}_{-1}$ the conditional distribution of negative examples. Then, we can analyze the generalization performance on a training set of size m containing m^+ positive examples and m^- negative examples by assuming that $S \sim \mathcal{D}_{+1}^{m^+} \otimes \mathcal{D}_{-1}^{m^-}$.

Under this assumption, the training set $S = ((x_1, y_1), ..., (x_m, y_m))$ satisfies $y_1 = y_2 = ... = y_{m^+} = 1$ and $y_{m^+ + 1} = ... = y_m = -1$. The empirical AUC risk (4.5) is given by:

$$E^{AUC}(f, S) = \frac{1}{m^+ m^-} \sum_{i: y_i = 1} \sum_{j: y_j = -1} \mathbb{1}_{[f(x_i) \leq f(x_j)]},$$

and is equal to $E^{\mathfrak{T}}(\tilde{f}, S)$, where $\tilde{f}$ is the classifier of pairs associated to f given by (4.7) and where $\mathfrak{T}(S)$ creates $m^+ m^-$ pairs out of m^+ positive examples and m^- negative examples as follows:

$$\mathfrak{T}(S) = ((1, (x_1, x_{m^+ + 1}), 1), ..., (1, (x_1, x_m), 1),$$

$$(1, (x_2, x_{m^+ + 1}), 1), ..., (1, (x_2, x_m), 1),$$

$$... \tag{4.9}$$

$$(1, (x_{m^+}, x_{m^+ + 1}), 1), ..., (1, (x_{m^+}, x_m), 1)).$$

The assumption $S \sim \mathcal{D}_{+1}^{m^+} \otimes \mathcal{D}_{-1}^{m^-}$ guarantees that

$$\mathcal{E}^{\mathfrak{T}}(\tilde{f}) = \mathbb{E}_{S \sim \mathcal{D}_{+1}^{m^+} \otimes \mathcal{D}_{-1}^{m^-}} [E^{\mathfrak{T}}(\tilde{f}, S)] = \mathcal{E}^{AUC}(f).$$

EXAMPLE **Ranking of Alternatives: Multiclass Classification**
The conditional pairwise comparison risk in ranking of alternatives can also
be written in our framework. We give here as example the simplest case of
multiclass classification given by (4.6):

$$E^{mc}(f,S) = \frac{1}{m(K-1)} \sum_{i=1}^{m} \sum_{k=2}^{K} \mathbb{1}_{\left[f(x_i^{\mu(\mathbf{y}_i,1)}) \leq f(x_i^{\mu(\mathbf{y}_i,k)}) \right]}.$$

Here the training set $S = ((\mathbf{x}_1, \mathbf{y}_1), ..., (\mathbf{x}_m, \mathbf{y}_m))$ is assumed to contain i.i.d.
samples from $\mathcal{D}$. It can be immediately seen that E^{mc} is then $E^{\mathfrak{T}}$ (4.8) with the
transformation that generates $m(K-1)$ pairs as follows:

$$\mathfrak{T}(S) = ((1, (x_1^{\mu(\mathbf{y}_i,1)}, x_1^{\mu(\mathbf{y}_i,2)}), 1), ..., (1, (x_1^{\mu(\mathbf{y}_i,1)}, x_1^{\mu(\mathbf{y}_i,K)}), 1)$$

$$...$$ (4.10)

$$(1, (x_m^{\mu(\mathbf{y}_i,1)}, x_m^{\mu(\mathbf{y}_i,2)}), 1), ..., (1, (x_m^{\mu(\mathbf{y}_i,1)}, x_m^{\mu(\mathbf{y}_i,K)}), 1)).$$

4.2.2   *Janson's Theorem and Interpretation*

We first describe here the result of Janson (2004), presented in Theorem 4.1 in an
abstract way, and make the link with the classification of pairs in the next subsection.
The description we use here, based on a dependency graph between random variables,
is a slightly simplified version of Janson's original formulation, but it is nonetheless
sufficient for our purposes.

Janson's result considers a sequence $(Z_1, ..., Z_M)$ of random variable that follow a *known* dependency structure that can be represented as an undirected graph
$\mathcal{G} = (\mathcal{V}, \mathcal{A})$ with vertices $\mathcal{V} = \{1, ..., M\}$ and where each random variable Z_i is
independent of the family $\{Z_j : (i, j) \notin \mathcal{A}\}$. The theorem gives a concentration
inequality for the sum $\sum_{i=1}^{M} Z_i$ assuming that the Z_is are bounded, in a similar
way to Hoeffding's inequality. To give first concrete examples, it is clear that if
$Z_1, ..., Z_m$ are independent, then $\mathcal{A}$, while if $Z_1, ..., Z_m$ are jointly distributed without
any independent subset, then $\mathcal{G}$ is the complete graph.

The structural feature of the dependency graph that is of use in Theorem 4.1 is the
notion of fractional chromatic number $\chi^*(\mathcal{G})$ of the graph $\mathcal{G}$. This number is based
on fractional covers of the vertices in the graph and is a refinement of the chromatic
number, which itself is based on covers of $\mathcal{V}$ with sets of non-adjacent vertices. Even
though these are general concepts from graph theory, we remind now their formal
definitions before stating the result; more details can be found in Janson's original
article.

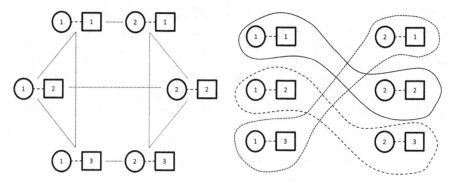

Fig. 4.3 (*left*) Dependency graph between the pairs generated for a bipartite ranking problem with two positive examples (*circles*) and three negative examples (*squares*). (*right*) Description of a cover of the vertices of the dependency graph into 3 subsets containing independent pairs

A *cover* of size K of G is a set $\{\mathfrak{M}_1, ..., \mathfrak{M}_K\}$ where each $\mathfrak{M}_k$ is a subset of V and $\bigcup_k \mathfrak{M}_k = V$ with the additional restriction that any two vertices in the same subset $\mathfrak{M}_k$ are non-adjacent, or, equivalently:

$$\bigcup_k \mathfrak{M}_k = V \text{ and } \{i, j\} \subset \mathfrak{M}_k \Rightarrow (i, j) \notin \mathcal{A}.$$

The *chromatic number* of $\mathcal{G}$, denoted by $\chi(\mathcal{G})$ is the smallest K such that there exists a cover of size K of $\mathcal{G}$. As a side note, the chromatic number is the quantity used in graph coloring problems, and is the minimum number of colors needed when we have to assign a color to each vertex under the constraint that no two adjacent vertices can be assigned the same color. For instance, when $\mathcal{A}$ is empty (i.e. the Z_i are independent), then $\chi(\mathcal{A}) = 1$, while for the complete graph $\chi(\mathcal{A}) = M$.

Figure 4.3 (left) contains the dependency graph between pairs for the example of bipartite ranking previously represented in Fig. 4.2. Since each pair contains one positive and one negative example, there is an edge between two pairs if and only if the two pairs have the same positive example or the same negative example. Figure 4.3 (right) represents a cover of the vertices of the graph into independent subsets (i.e. subsets containing only non-adjacent vertices).

As a refinement to the notion of cover, a *fractional cover* of $\mathcal{G}$ is a set of couples $\{(\omega_1, \mathfrak{M}_1), ..., (\omega_K, \mathfrak{M}_K)\}$ where each $\omega_k \geq 0$ and $\mathfrak{M}_K \subset V$ such that:

1. $\{\mathfrak{M}_1, ..., \mathfrak{M}_K\}$ is a cover of $\mathcal{G}$ with sets of non-adjacent vertices, and
2. $\forall i \in V, \sum_{k=1}^K \omega_k \, \mathbb{1}_{[i \in \mathfrak{M}_k]} \geq 1$. A fractional cover is called proper if for all i, the inequality is an equality. *Throughout this work, we only consider proper fractional covers.*

In other words, each set $\mathfrak{M}_k$ contains only non-adjacent vertices, and has a weight ω_k attached to it. A single vertex might appear in one or several of the ω_k, but their is the restriction that the weights of the sets in which a vertex appears must sum to at least 1. The *fractional chormatic number* of $\mathcal{G}$, denoted by $\chi^*(\mathcal{G})$ is the minimum value of $\sum_{k=1}^{K} \omega_k$ over all possible fractional covers $\{(\omega_1, \mathfrak{M}_1), ..., (\omega_K, \mathfrak{M}_K)\}$ of $\mathcal{G}$.

By definition, the fractional chromatic number is smaller or equal to the chromatic number, which itself is smaller than 1 plus the maximum degree of $\mathcal{G}$. In graph coloring problems, the fractional chromatic number of a graph if the infimum of a/b with $a, b \in \mathbb{N}$ such that we can assign to each vertex of $\mathcal{G}$ exactly b out of a possible colors with the constraint that any two adjacent vertices have no colors in common.

The major interest of these definitions is that if we have a (proper) fractional cover $((\omega_k, \mathfrak{M}_k))_{k=1}^{K}$, then the sum of interdependent variables can be decomposed into a weighted sum of sums of independent variables as follows:

$$\sum_{i=1}^{M} Z_i = \sum_{i=1}^{M} \sum_{k=1}^{K} \omega_k \, \mathbb{1}_{[i \in \mathfrak{M}_k]} Z_i = \sum_{k=1}^{K} \omega_k \sum_{i \in \mathfrak{M}_k} Z_i \tag{4.11}$$

We shall call this decomposition Janson's decomposition from now on.

Using these definitions, Janson's theorem can be stated as follows (Janson 2004):

Theorem 4.1 *Let* $(Z_1, ..., Z_M)$ *be a sequence of* M *random variable, and let* $\chi^*(\mathcal{G})$ *the fractional chromatic number of the dependency graph of* $\{Z_i\}_{i=1}^{M}$. *Assuming that* $\forall i, \exists (a_i, b_i) \in \mathbb{R}^2$ *such that* $Z_i \in [a_i, b_i]$ *we have:*

$$\forall \epsilon > 0, \quad \mathbb{P}\left(\mathbb{E}\left[\sum_{i=1}^{M} Z_i \right] - \sum_{i=1}^{M} Z_i > \epsilon \right) \leq \exp\left(-\frac{2\epsilon^2}{\chi^*(\mathcal{G}) \sum_{i=1}^{M} (b_i - a_i)^2} \right).$$

The bound we just presented is a strict extension of Hoeffding's inequality, since when the Z_is are independent, we have $\chi^*(\mathcal{G}) = 1$ and we recover exactly Hoeffding's inequality. The fractional chromatic number of the dependency graph is thus, in this analysis, a measure of how much the lack of independence between the random variables affects the concentration of the sum around its expectation.

Remark 4.3 In machine learning, it is often convenient to express concentration inequalities as a function of the desired probability instead of a desired deviation value ϵ.

Fixing $\delta = exp(-\frac{2\epsilon^2}{\chi^*(\mathcal{G}) \sum_{i=1}^{M} (b_i - a_i)^2})$ and solving for ϵ gives:

$$\epsilon = \sqrt{\chi^*(\mathcal{G}) \sum_{i=1}^{M} (b_i - a_i)^2 \ln(1/\delta)/2}.$$

By reverting the inequality inside the probability of Theorem 4.1, we obtain, under the same assumptions:

$$\forall \delta \in (0, 1], \mathbb{P}\left(\mathbb{E}\left[\sum_{i=1}^{M} Z_i \right] \leq \sum_{i=1}^{M} Z_i + \sqrt{\chi^*(\mathcal{G}) \sum_{i=1}^{M} (b_i - a_i)^2 \ln(1/\delta)/2} \right) \geq 1 - \delta.$$

$$\tag{4.12}$$

4.2.3 Generic Test Bounds

Theorem 4.1 can readily be used to derive test set bounds in the framework of classification described above. Indeed, given a transformation $\mathfrak{T}$ and a training set $S = ((x_i, y_i))_{i=1}^m$ of size m sampled according to $\mathcal{D}_1^m$, the dependency graph of the M random variables $\mathfrak{T}(S) = ((\tilde{z}_i, \tilde{x}_i, \tilde{y}_i))_{i=1}^M$ is the graph connecting i and j if at least one of the (x_i, y_i) is used in the computation of both $(\tilde{z}_i, \tilde{x}_i, \tilde{y}_i)$ and $(\tilde{z}_j, \tilde{x}_j, \tilde{y}_j)$.

Denoting $\chi^*(\mathfrak{T})$ the fractional chromatic number of the dependency graph of $\mathfrak{T}(S)$ when S contains independent samples, and assuming that $\tilde{f}$ is a fixed classifier over $\tilde{\mathcal{X}}$, Theorem 4.1 (and more precisely 4.12) applied to the random variables $Z_i = \frac{1}{M} \tilde{z}_i \, \mathbb{1}_{[\tilde{f}(\tilde{x}_i) \neq \tilde{y}_i]} \in [0, 1/M]$ immediately gives us:

$$\forall \delta \in (0, 1], \mathbb{P}_{S \sim \mathcal{D}_1^m} \left(\mathcal{E}^{\mathfrak{T}}(\tilde{f}) \leq E^{\mathfrak{T}}(\tilde{f}, S) + \sqrt{\frac{\chi^*(\mathfrak{T}) \ln(1/\delta)}{2M}} \right) \geq 1 - \delta. \quad (4.13)$$

Thus, we obtain a generic test set bound which only needs to value of $\chi^*(\mathfrak{T})$ to be computed in practice. Technically, it can be obtained by direct calculations given $\mathfrak{T}$. We give here the proof for the AUC risk (4.5) in bipartite ranking for which $M = m^+ m^-$ and $\chi^*(\mathfrak{T}) = \max(m^+, m^-)$ where m^+ and m^- are respectively the number of positive and negative examples in the training set:

Corollary 4.1 *Let $\mathcal{D}_{+1}$ and $\mathcal{D}_{-1}$ be the conditional distributions over $\mathcal{X} \times \{-1, 1\}$ of positive and negative examples respectively. Let $(m^+, m^-) \in (\mathbb{N} \backslash \{0\})^2$. Then, for all $\delta \in (0, 1]$, we have:*

$$\mathbb{P}_{S \sim \mathcal{D}_{+1}^{m^+} \otimes \mathcal{D}_{-1}^{m^-}} \left(\mathcal{E}^{AUC}(f) \leq E^{AUC}(f, S) + \sqrt{\frac{\ln(1/\delta)}{2 \min(m^+, m^-)}} \right) \geq 1 - \delta.$$

PROOF
Coming back to the example on the Area Under the ROC curve given in Sect. 4.2.1, we know that:
- $E^{AUC}(f, S) = E^{\mathfrak{T}}(\tilde{f}, S)$ if $\tilde{f}$ is taken as the classifiers of pairs associated to f, and $\mathfrak{T}$ is given by (4.9);
- $\mathcal{E}^{AUC}(f) = \mathbb{E}_{S \sim \mathcal{D}_{+1}^{m^+} \otimes \mathcal{D}_{-1}^{m^-}} E^{\mathfrak{T}}(\tilde{f}, S) = \mathcal{E}^{\mathfrak{T}} f$.

Applying the generic test set bound (4.13), we obtain the result up to $\frac{1}{\min(m^+, m^-)}$ replaced by $\frac{\chi^*(\mathfrak{T})}{m^+ m^-}$ (recall that $M = m^+ m^-$). We thus need to prove $\chi^*(\mathfrak{T}) \leq \max(m^+, m^-)$.

To that end, let us assume that $m^+ <= m^-$ (the other case is similar). We now prove that we can obtain a cover $(\mathfrak{M}_1, ..., \mathfrak{M}_{m^+})$ of size m^- of the $m^+ m^-$ pairs created by $\mathfrak{T}$ where each $\mathfrak{M}_k$ only contains independent pairs. To simplify notation, we denote $p_{i,j} = (1, (x_i, x_j), 1)$, i.e. the transformed example created by $\mathfrak{T}$ containing the pair of inputs (x_i, x_j), and we remind that if $S = ((x_1, y_1), ..., (x_m, y_m))$, then the m^+ first examples are positive and the m^- remaining ones are negative. The cover is constructed using the "roundrobin" pattern represented in Fig. 4.3 (right) and which can be generalized as follows:

$$\mathfrak{M}_1 = (p_{1,m^++1}, p_{2,m^++2}, ..., p_{m^+,m^++m^+})$$
$$\mathfrak{M}_2 = (p_{1,m^++2}, p_{2,m^++3}, ..., p_{m^+,m^++m^++1})$$

$$...$$

$$\mathfrak{M}_{m^--m^++1} = (p_{1,m^-+1}, p_{2,m^-+2}, ..., p_{m^+,m})$$
$$\mathfrak{M}_{m^--m^++2} = (p_{1,m^-+2}, p_{2,m^-+3}, ..., p_{m^+,m^++1})$$

$$...$$

$$\mathfrak{M}_{m^-} = (p_{1,m}, p_{2,m^++1}, ..., p_{m^+,m^++m^+-1}).$$

That is, we build each $\mathfrak{M}_k$ by taking the m^+ positive examples and m^+ distinct negative examples, and assign a different negative example to each positive example in order to make sure all the pairs in $\mathfrak{M}_k$ are independent. The round-robin pattern makes sure that all the pairs are covered with m^- such sets. This cover ensures $\chi(\mathfrak{T}) \leq m^- = \max(m^+, m^-)$ and thus $\chi^*(\mathfrak{T}) = \max(m^+, m^-)$.

The bound given in Corollary 4.1 is strictly better than the previous bound of Agarwal et al. (2005) that was based on McDiarmid's Theorem: The bound of Agarwal et al. has a factor $\frac{1}{m^+} + \frac{1}{m^-}$ instead of our $\frac{1}{\min(n_+, n_-)}$, which can be larger by a factor of 2. This shows that our method for dealing with interdependent data using Janson's approach can lead to tighter bounds than previous approaches. It is, also, more general to some extent since the same bound applies to many problems, up to the computation of $\chi^*(\mathfrak{T})$.

To give more examples, it can easily be checked on the examples of Sect. 4.2.1 that $\chi^*(\mathfrak{T}) \leq m$ for the pairwise comparison risk in the general case of ranking of instances. We will see in more details in Sect. 4.3.5 that $\chi^*(\mathfrak{T}) = K - 1$ for the conditional pairwise comparison risk in multiclass classification.

4.3 Generalization Bounds for Learning with Interdependent Data

We described in the last section the core of our framework, as well as the main tool underlying our analysis: the transformation of the examples, and the application of Janson's decomposition to the dependency graph of the transformed examples to obtain concentration inequalities.

In this section, we present how these tools can be leveraged to extend the Rademacher complexity analysis presented in Sect. 2.3 to the case of interdependent data. The main steps and tools to obtain the Rademacher complexity analysis are McDiarmid's Theorem (McDiarmid 1989) and the symmetrization step. In this section, we extend both of these results to the case of interdependent data, using Janson's decomposition of the dependency graph based on fractional covers. We first start with an extension of both McDiarmid's Theorem and Janson inequality (Theorem 4.1) to more general functions than sums. We then describe how the symmetrization step and the notion of Rademacher complexity can be extended to the case of interdependent data.

4.3.1 Extension of McDiarmid's Theorem

The result we present in this and the remaining sections assumes we have:

- a sequence of m *independent* random variables $S = (X_1, ..., X_m) \sim \mathcal{D}_1^m$, where each X_i takes values in the input space $\mathcal{X}$,
- a transformation function $\mathfrak{T}(S) : \mathcal{X}^m \to \tilde{\mathcal{X}}^M$, and,
- an optimal fractional cover $((\omega_1, \mathfrak{M}_1), ..., (\omega_K, \mathfrak{M}_K))$ of the dependency graph of $\mathfrak{T}(S)$, where each $\mathfrak{M}_k$ contains independent random variables (see Sect. 4.2.2). We assume the fractional cover is optimal in the sense that $\sum_{k=1}^{K} \omega_k = \chi^*(\mathfrak{T})$.

Moreover, for notational convenience, we denote by M_k the cardinal of $\mathfrak{M}_k$ and use an index for the integers of $\mathfrak{M}_k = \{n_{k,1}, ..., n_{k,M_k}\}$.

The extension of McDiarmid's theorem we propose is the following ((Usunier et al. 2006), Theorem 2):

Theorem 4.2 *Under the notation described above, let $\varphi : \tilde{\mathcal{X}}^M \to \mathbb{R}$ and, for $k \in \{1, ..., K\}$ let $\varphi^k : \tilde{\mathcal{X}}^{M_k} \to \mathbb{R}$ such that:*

1. *$\forall(z_1, ..., z_m) \in \tilde{\mathcal{X}}^M, \varphi(z_1, ..., z_m) = \sum_{i=1}^{K} \omega_k \varphi^k(z_{n_{k,1}}, ..., z_{n_{k,M_k}})$,*
2. *the φ^ks have bounded differences: $\exists c_1 \in \mathbb{R}, ..., c_M \in \mathbb{R}$ such that:*

$$\forall k \in \{1, ..., K\}, \forall i \in \{1, ..., M_k\}, \forall(z_1, ..., z_m) \in \tilde{\mathcal{X}}^M, \forall z \in \tilde{\mathcal{X}} :$$

$$|\varphi^k(z_{n_{k,1}}, ..., z_{n_{k,M_k}}) - \varphi^k(z_{n_{k,1}}, ..., z_{n_{k,i-1}}, z, z_{n_{k,i+1}} ..., z_{n_{k,M_k}})| \leq c_{n_{k,i}}.$$

Then, for all $\epsilon > 0$:

$$\mathbb{P}_{S \sim \mathcal{D}_1^m} \left(\varphi(\mathfrak{T}(S)) - \mathbb{E}_{S' \sim \mathcal{D}_1^m}[\varphi(\mathfrak{T}(S'))] > \epsilon \right) \leq \exp \left(\frac{-2\epsilon^2}{\chi^*(\mathfrak{T}) \sum_{i=1}^{M} c_i^2} \right),$$

and:

$$\mathbb{P}_{S \sim \mathcal{D}_1^m} \left(\mathbb{E}_{S' \sim \mathcal{D}_1^m}[\varphi(\mathfrak{T}(S'))] - \varphi(\mathfrak{T}(S)) > \epsilon \right) \leq \exp \left(\frac{-2\epsilon^2}{\chi^*(\mathfrak{T}) \sum_{i=1}^{M} c_i^2} \right),$$

The theorem above is a strict extension of McDiarmid's inequality, since the latter corresponds to the case where the transformation $\mathfrak{T}$ is the identity function of $\mathcal{X}^m$ with the cover of size 1 defined by $(1, \{1, ..., M\})$ and $\varphi^1 = \varphi$.

The theorem above is also a strict extension of Janson's result (Theorem 4.1), since if one takes $\tilde{\mathcal{X}} = \mathbb{R}$ φ and all the φ^k as being the sums of their arguments, the first condition corresponds to Janson's decomposition (4.11), and the deviation bound that is finally obtained is the same as in Theorem 4.1.

PROOF **of Theorem 4.2**

The proof essentially follows that of Janson (2004), plugging a lemma from the proof of McDiarmid's theorem at the critical moment.

In order to simply notation, we will use a functional notation:

$$\mathbb{E}[\varphi \circ \mathfrak{T}] = \mathbb{E}_{S' \sim \mathcal{D}_1^m}[\varphi(\mathfrak{T}(S'))] \qquad (4.14)$$

and we will denote $\mathfrak{T}^k(S) = (Z_{n_{k,1}}, ..., Z_{n_{k,M_k}})$ where $(Z_1, ..., Z_n) = \mathfrak{T}(S)$ and use the shorter notation $\mathbb{E}[\varphi^k \circ \mathfrak{T}^k]$ defined analogously as in (4.14). Moreover, all probabilities and expectations are implicitly assumed to be taken according to $S \sim \mathcal{D}_1^m$.

We now only prove the first inequality, the second inequality follows by taking $\varphi = -\varphi$.

First, following Hoeffding's method (Hoeffding 1963), we have, for any $s > 0$ and $t > 0$:

$$\mathbb{P}(\varphi \circ \mathfrak{T} - \mathbb{E}[\varphi \circ \mathfrak{T}] > t) \leq e^{-st} \mathbb{E}[e^{s(\varphi \circ \mathfrak{T} - \mathbb{E}[\varphi \circ \mathfrak{T}])}]$$

$$= e^{-st} \mathbb{E}\left[exp\left(s\left(\sum_{k=1}^{K} \omega_k(\varphi^k \circ \mathfrak{T}^k - \mathbb{E}\left[\varphi^k \circ \mathfrak{T}^k\right])\right)\right)\right].$$

Now, let $\beta_1, ..., \beta_K$ be any set of K strictly positive reals that sum to 1. Since $\sum_{k=1}^{K} \omega_k / \chi^*(\mathfrak{T}) = 1$, using the convexity of the exponential and Jensen's inequality, we obtain:

$$e^{-st} \mathbb{E}\left[exp\left(s\left(\sum_{k=1}^{K} \omega_k(\varphi^k \circ \mathfrak{T}^k - \mathbb{E}\left[\varphi^k \circ \mathfrak{T}^k\right])\right)\right)\right]$$

$$\leq e^{-st} \sum_{k=1}^{K} \beta_k \mathbb{E}\left[exp\left(\frac{s\omega_k}{\beta_k}(\varphi^k \circ \mathfrak{T}^k - \mathbb{E}\left[\varphi^k \circ \mathfrak{T}^k\right])\right)\right]. \quad (4.15)$$

We can now apply the following lemma, which appears in the proof of McDiarmid's inequality (McDiarmid 1989), which we can apply to each term of the main sum because these term consider independent subsets of $\mathfrak{T}(\mathfrak{T})$:

Lemma 4.1 *Let $X_1, ..., X_m$, m independent random variables taking values in $\mathcal{X}$ and $\varphi : \mathcal{X}^m \to \mathbb{R}$ be a function with differences bounded by c_i on dimension i. Then:*

$$\mathbb{E}[e^{s(\varphi(X_1,...,X_m)-\mathbb{E}[\varphi])}] \leq e^{s^2/8 \sum_{i=1}^{m} c_i^2}$$

Applying this lemma to each term in the sum of (4.15), we obtain:

$$\mathbb{P}(\varphi \circ \mathfrak{T} - \mathbb{E}[\varphi \circ \mathfrak{T}] > t) \leq e^{-st} \sum_{k=1}^{K} \beta_k exp \left(\frac{(s\chi^*(\mathfrak{T}))^2}{8\beta_k^2} \sum_{i=1}^{M_k} c_{n_k,i}^2 \right).$$

The right-hand term is the same to the one appearing in Janson's theorem proof, so that the optimal values are the same. Let us define:

- $C_k = \sum_{i \in \mathfrak{m}_k} c_i^2$,
- $C = \sum_{k=1}^{K} \omega_k \sqrt{C_k}$,
- $b_k = \omega_k \sqrt{C_k}/C$,
- $s = 4t/C^2$,

then, we can check that the last equation gives:

$$\mathbb{P}(\varphi \circ \mathfrak{T} - \mathbb{E}[\varphi \circ \mathfrak{T}] > t) \leq e^{-2t^2/C^2}.$$

The end of the proof is given by Cauchy-Schwarz inequality:

$$C^2 = \left(\sum_{k=1}^{K} \omega_k \sqrt{C_k} \right)^2 \leq \left(\sum_{k=1}^{K} \omega_k \right) \left(\sum_{k=1}^{K} \omega_k C_k \right) = \chi^*(\mathfrak{T}) \sum_{i=1}^{M} c_i^2.$$

4.3.2 The Fractional Rademacher Complexity

Now we have an extension of McDiarmid's theorem to deal with interdependent data, we are now ready to extend the Rademacher complexity analysis. For now, we only considered errors for learning with interdependent data of the form given by (4.8) and (4.8):

$$\mathcal{E}^{\mathfrak{T}}(\tilde{f}) = \mathbb{E}_{S \sim \mathcal{D}_1^m}[E^{\mathfrak{T}}(\tilde{f}, S)] \quad \text{where} \quad E^{\mathfrak{T}}(\tilde{f}, S) = \frac{1}{M} \sum_{i=1}^{M} \tilde{z}_i \mathbb{1}_{[\tilde{f}(\tilde{x}_i) \neq \tilde{y}_i]}.$$

The Rademacher complexity is however a more general notion that can by applied to many other definitions of risk. In order to make the analysis more general, but still sticking to our framework, let us consider a class of functions $\mathcal{F}$ and a fixed instantaneous loss function $L : \mathcal{F} \times (\{0, 1\} \times \tilde{\mathcal{X}} \times \{-1, 1\}) \to \mathbb{R}$. We do not make any restriction on the domain of the functions in $\mathcal{F}$, we simply assume that L can take any function in $\mathcal{F}$, any transformed example and gives some measure of performance. Then, define the empirical L-risk with interdependent data as:

$$E_L^{\mathfrak{T}}(f, S) = \frac{1}{M} \sum_{i=1}^{M} L(f, \tau_i)$$

$$\text{where } (\tau_1, ..., \tau_M) = ((\tilde{z}_1, \tilde{x}_1, \tilde{y}_1), ..., (\tilde{z}_M, \tilde{x}_M, \tilde{y}_M)) = \mathfrak{T}(S) \qquad (4.16)$$

so that we define the L-risk as the expectation of $E_L^{\mathfrak{T}}$:

$$\mathcal{E}_L^{\mathfrak{T}}(f, S) = \mathbb{E}_{S \sim \mathcal{D}_1^m}[E_L^{\mathfrak{T}}(f, S)]. \qquad (4.17)$$

This slight generalization of the instantaneous loss is somewhat necessary to be able to consider other losses than the 0/1-classification error, such as margin-based losses of (Ranking) Support Vector Machines.

Now, the extension of the Rademacher complexity we propose is based on the following result, which is the premise to the symmetrization step and the first application of the extension of McDiarmid's theorem (Theorem 4.2):

Lemma 4.2 *Using the notation above, let additionally $((\omega_k, \mathfrak{M}_k))_{k=1}^{K}$ be a fractional cover of independent set for $\mathfrak{T}$ with $\sum_{k=1}^{K} \omega_k = \chi^*(\mathfrak{T})$, and let us assume that $||L||_\infty \leq B$. Then, for all $\delta \in (0, 1]$, with probability at least $1 - \delta$ over training sets S sampled according to $\mathcal{D}_1^m$, we have:*

$$\sup_{f \in \mathcal{F}} (\mathcal{E}_L^{\mathfrak{T}}(f) - E_L^{\mathfrak{T}}(f, S))$$

$$\leq \sum_{k=1}^{M} \frac{\omega_k}{M} \mathbb{E}_{S' \sim \mathcal{D}_1^m, S'' \sim \mathcal{D}_1^m} \left[\sup_{f \in \mathcal{F}} \sum_{i \in \mathfrak{M}_k} (L(f, \tau_i') - L(f, \tau_i'')) \right]$$

$$+ B \sqrt{\frac{\chi^*(\mathfrak{T}) \ln (1/\delta)}{2M}}.$$

where we use the notation $(\tau_1', ..., \tau_M') = \mathfrak{T}(S')$ and $(\tau_1'', ..., \tau_M'') = \mathfrak{T}(S'')$.

PROOF
Starting from the definition of $\mathcal{E}_L^{\mathfrak{T}}$, we have:

$$\sup_{f \in \mathcal{F}} (\mathcal{E}_L^{\mathfrak{T}}(f) - E_L^{\mathfrak{T}}(f, S)) = \sup_{f \in \mathcal{F}} (\mathbb{E}_{S' \sim \mathcal{D}_1^M}[E_L^{\mathfrak{T}}(f, S')] - E_L^{\mathfrak{T}}(f, S))$$

$$\leq \mathbb{E}_{S' \sim \mathcal{D}_1^M} \left[\sup_{f \in \mathcal{F}} (E_L^{\mathfrak{T}}(f, S') - E_L^{\mathfrak{T}}(f, S)) \right]$$

because the supremum of the expected value is smaller than the expected value of the supremum. Now, using Janson's decomposition (4.11), we have

$$E_L^{\mathfrak{T}}(f, S') - E_L^{\mathfrak{T}}(f, S) = \sum_{k=1}^{K} \omega_k \sum_{i \in \mathfrak{M}_k} (L(f, \tau_i') - L(f, \tau_i)).$$

Using the superadditivity of the supremum and the linearity of the expectation, we obtain:

$$\sup_{f \in \mathcal{F}} (\mathcal{E}_L^{\mathfrak{T}}(f) - E_L^{\mathfrak{T}}(f, S)) \leq \sum_{k=1}^{M} \frac{\omega_k}{M} \mathbb{E}_{S' \sim \mathcal{D}_1^m} \left[\sup_{f \in \mathcal{F}} \sum_{i \in \mathfrak{M}_k} (L(f, \tau_i') - L(f, \tau_i)) \right].$$

Now, defining $\varphi(\tau_1, ... \tau_M)$ being equal to the right-hand side of the equation, we can apply Theorem 4.2 to φ using

$$\varphi^k(\tau_{n_{k,1}}, ..., \tau_{n_{k,M_k}}) = \frac{1}{M} \mathbb{E}_{S' \sim \mathcal{D}_1^m} \left[\sup_{f \in \mathcal{F}} \sum_{i=1}^{M_k} \left(L(f, \tau_{n_{k,i}}') - L(f, \tau_{n_{k,i}}) \right) \right],$$

since all φ^k have differences (in the sense of condition 2 of Theorem 4.2 bounded by $\frac{B}{M}$, we obtain the desired result.

Lemma 4.2 provides us with the insight that an extension of the Rademacher complexity that can handle interdependent examples can be found using a decomposition into independent sets. The notion of *fractional Rademacher complexity*, which we define below, extends the Rademacher complexity in such a way (Usunier et al. (2006), Definition 3):

Definition 4.1 (Fractional Rademacher complexity) Let $((\omega_k, \mathfrak{M}_k))_{k=1}^{K}$ be a fractional cover of independent sets for $\mathfrak{T}$ with $\sum_{k=1}^{K} \omega_k = \chi^*(\mathfrak{T})$ and let:

$$L \circ \mathcal{F} = \{\tau \mapsto L(f, \tau) | f \in \mathcal{F}\}.$$

Then, the *fractional Rademacher complexity* of $L \circ \mathcal{F}$ is defined by:

$$\mathcal{R}_m^{\mathfrak{T}}(L \circ \mathcal{F}) = \sum_{k=1}^{K} \frac{2\omega_k}{M} \mathbb{E}_{S \sim \mathcal{D}_1^m} \mathbb{E}_{\sigma} \left[\sup_{f \in \mathcal{F}} \left| \sum_{i \in \mathfrak{M}_k} \sigma_i L(f, \tau_i) \right| \right],$$

where $\sigma \in \{-1, 1\}$ is a vector of M independent Rademacher variables (i.e. $\mathbb{P}(\sigma_i = 1) = \mathbb{P}(\sigma_i = -1) = \frac{1}{2}$).

Moreover, the *empirical fractional Rademacher complexity* given a training set S is defined by:

$$R^{\mathfrak{T}}(L \circ \mathcal{F}, S) = \sum_{k=1}^{K} \frac{2\omega_k}{M} \mathbb{E}_\sigma \left[\sup_{f \in \mathcal{F}} \left| \sum_{i \in \mathfrak{M}_k} \sigma_i L(f, \tau_i) \right| \right].$$

The fractional Rademacher complexity is a strict extension of the Rademacher complexity since the latter corresponds to the case where the transformation $\mathfrak{T}$ is the identity function.

We may also notice that the fractional Rademacher complexity is defined through a reference to a particular fractional cover. In general, our results hold for any optimal fractional cover; different choices of covers might lead to different values of the bounds. Nonetheless, this effect is unlikely to have important impacts in practice.

Using Definition 4.1, we can perform the symmetrization step and state the main result (Usunier et al. (2006), Theorem 4):

Theorem 4.3 *Assume that S contains m independent samples drawn according to $\mathcal{D}_1^m$, and let $B = ||L||_\infty$. Then, for any $\delta \in (0, 1]$, with probability at least $1 - \delta$, the two following bounds hold for all $f \in \mathcal{F}$:*

$$\mathcal{E}_L^{\mathfrak{T}}(f) \leq E_L^{\mathfrak{T}}(f, S) + \mathcal{R}_m^{\mathfrak{T}}(L \circ \mathcal{F}) + B \sqrt{\frac{\chi^*(\mathfrak{T}) \ln(1/\delta)}{2M}},$$

$$\mathcal{E}_L^{\mathfrak{T}}(f) \leq E_L^{\mathfrak{T}}(f, S) + R^{\mathfrak{T}}(L \circ \mathcal{F}, S) + 3B \sqrt{\frac{\chi^*(\mathfrak{T}) \ln(2/\delta)}{2M}}.$$

The bounds obtained in this theorem are exactly the same as the usual Rademacher complexity bounds presented in Theorem 2.2 of Chap. 2, but where the usual risk assuming i.i.d. examples is replaced by the risk considering interdependent examples, and the Rademacher complexity replaced by the fractional Rademacher complexity. In the case where the transformed examples are i.i.d., the two bounds are equivalent.

As we have seen in the generic test set bound for interdependent examples of Sect. 4.2.3, the concentration inequalities are similar to those using independent samples, up to the fact that the number of examples M has to be replaced by $\frac{M}{\chi^*(\mathfrak{T})}$. In the next section, we make the link between the usual Rademacher complexity and the fractional Rademacher complexity, and show that these two quantities are also similar up to the factor $\chi^*(\mathfrak{T})$ in the number of examples.

PROOF **of Theorem 4.3**
Starting from Lemma 4.2, we need to notice that if $\sigma = (\sigma_1, ..., \sigma_M) \in \{-1, 1\}^M$, we claim that:

$$\mathbb{E}_{S' \sim \mathcal{D}_1^m, S'' \sim \mathcal{D}_1^m} \left[\sup_{f \in \mathcal{F}} \sum_{i \in \mathfrak{M}_k} \left(L(f, \tau_i') - L(f, \tau_i'') \right) \right]$$

$$= \mathbb{E}_{S' \sim \mathcal{D}_1^m, S'' \sim \mathcal{D}_1^m} \left[\sup_{f \in \mathcal{F}} \sum_{i \in \mathfrak{M}_k} \sigma_i \left(L(f, \tau_i') - L(f, \tau_i'') \right) \right].$$

Indeed, if $\sigma_i = -1$, then it is equivalent to compute the supremum with τ_i'' in $\mathfrak{T}(S')$ and τ_i' in $\mathfrak{T}(S')$. This, in turn, is equivalent to exchange between S' and S'' all the original examples of which τ_i' and τ_i'' depend. Since the we only consider the joint expectation over S' and S'', this exchange does not change the final value. Finally, all the values of σ_i can be arbitrary because we only consider a sum over an independent subset of transformed examples, so that exchange between S' and S'' performed by a single σ_i has no effect on the distribution of the other τ_j' and τ_j'' for $j \in \mathfrak{M}_k$.

Taking the expectation over σ of M independent Rademacher variables, we obtain:

$$\mathbb{E}_{S' \sim \mathcal{D}_1^m, S'' \sim \mathcal{D}_1^m} \left[\sup_{f \in \mathcal{F}} \sum_{i \in \mathfrak{M}_k} (L(f, \tau_i') - L(f, \tau_i'')) \right]$$

$$= \mathbb{E}_{S' \sim \mathcal{D}_1^m, S'' \sim \mathcal{D}_1^m} \mathbb{E}_\sigma \left[\sup_{f \in \mathcal{F}} \sum_{i \in \mathfrak{M}_k} \sigma_i (L(f, \tau_i') - L(f, \tau_i'')) \right].$$

$$\leq 2 \mathbb{E}_{S' \sim \mathcal{D}_1^m} \mathbb{E}_\sigma \left[\sup_{f \in \mathcal{F}} \left| \sum_{i \in \mathfrak{M}_k} \sigma_i L(f, \tau_i') \right| \right].$$

where the last inequality is obtained by the superadditivity of the supremum. The first inequality of the theorem directly follows by plugging that equation to Lemma 4.2 and using th definition of the fractional Rademacher complexity.

The second inequality of the theorem is obtained by applying the extension of McDiarmid's theorem to the fractional Rademacher complexity to obtain a bound using its empirical version. This corresponds to the third step to obtain a Rademacher complexity bound described in Chap. 2, Sect. 2.3. In our case, the result is obtained by applying Theorem 4.2 with:

$$\varphi^k(\tau_{n_{k,1}}, \ldots, \tau_{n_{k,M_k}}) = \frac{2}{M} \mathbb{E}_\sigma \left[\sup_{f \in \mathcal{F}} \left| \sum_{i=1}^{M_k} \sigma_i L(f, \tau_{n_{k,i}}) \right| \right].$$

4.3.3 Estimation of the Fractional Rademacher Complexity

The fractional Rademacher complexity is, in essence, a weighted sum of Rademacher complexities computed with independent samples. In this section, we show how the well-known properties of the Rademacher complexity described allow us to obtain estimates on the fractional Rademacher complexity.

The following result makes the link between the fractional Rademacher complexity and the VC dimension, as well as an upper bound in the case of kernel classes of function. This result is thus an extension of Corollary 2.2 and Eq. (2.33) of Chap. 2.

Theorem 4.4

1. *Assume that $L \circ \mathcal{F}$ is a class of binary functions on the space of transformed examples $\{0, 1\} \times \tilde{\mathcal{X}} \times \{-1, 1\}$, and that the VC dimension of $L \circ \mathcal{F}$, denoted by V, is finite.*
   Also assume that in each of the independent subsets $\mathfrak{M}_k$ of the fractional cover used to define the fractional Rademacher complexity, we have $M_k \geq V$. Then:

$$\mathcal{R}_m^{\mathfrak{T}}(L \circ \mathcal{F}) \leq \sqrt{8 \left(\frac{\chi^*(\mathfrak{T}) \ln 2}{M} + \frac{V \chi^*(\mathfrak{T})}{M} \ln \frac{eM}{V \chi^*(\mathfrak{T})} \right)}$$

2. *Let $\tilde{\mathcal{X}} = \mathbb{R}^d$ for some d, and assume that*

$$L \circ \mathcal{F} = \{(\tilde{z}, \tilde{x}, \tilde{y}) \mapsto \langle w\tilde{x} \rangle \mid \|w\|_2 \leq B\}$$

Then:

$$R^{\mathfrak{T}}(L \circ \mathcal{F}, S) = \frac{2B \chi^*(\mathfrak{T})}{M} \sqrt{\sum_{i=1}^m \|\tilde{x}_i\|_2^2}.$$

This result shows that, as expected, the interdependency between examples only affects the rate at which the complexity term converges to 0 by a factor $\chi^*(\mathfrak{T})$, but appart from that, the complexity measures used in classification and more generally in the context of learning with i.i.d. samples are fully relevant in the context of learning with interdependent samples.

For instance, the Rademacher complexity for linear classes of linear functions is widely used as a complexity measures for kernel machines such as Support Vector Machines (Bartlett and Mendelson 2003a; Taylor and Cristianini 2004). The second part of the theorem can thus be seen as an extension of these results to the cases of SVMs for ranking based on pairwise comparisons such as Weston and Watkins (1999); Herbrich et al. (1998); Joachims (2002b) (see also Sect. 4.1.3).

In terms of practical applications, the result essentially shows that the methods used for capacity control in the usual binary classification setup, such as regularization in support vector machines and structural risk minimization, can be readily applied with similar guarantees to problems of learning with interdependent data such as the various ranking problems. The analysis carried out in this chapter thus allows us to give guarantees on the statistical performance of the algorithms for learning to rank within a unified framework. We give some concrete examples in the next subsection, after proving Theorem 4.4.

PROOF **of Theorem 4.4**

1. Considering the definition of the fractional Rademacher complexity, we first explicitly write it as a weighted sum of (expectations of) empirical Rademacher complexities. Then, we bound the inner terms using the growth function approach, as in the proof of Corollary 2.2:

$$
\mathcal{R}_m^{\mathfrak{T}}(L \circ \mathcal{F}) = \sum_{k=1}^{K} \frac{2\omega_k}{M} \mathbb{E}_{S \sim \mathcal{D}_1^m} \mathbb{E}_\sigma \left[\sup_{f \in \mathcal{F}} \left| \sum_{i \in \mathfrak{M}_k} \sigma_i L(f, \tau_i) \right| \right]
$$

$$
\leq \sum_{k=1}^{K} \frac{M_k \omega_k}{M} \mathbb{E}_{S \sim \mathcal{D}_1^m} \left[\frac{2}{M_k} \mathbb{E}_\sigma \left[\sup_{f \in \mathcal{F}} \left| \sum_{i \in \mathfrak{M}_k} \sigma_i L(f, \tau_i) \right| \right] \right]
$$

$$
\leq \sum_{k=1}^{K} \frac{M_k \omega_k}{M} \mathbb{E}_{S \sim \mathcal{D}_1^m} \left[\sqrt{\frac{8 \ln (2 \mathfrak{G}(L \circ \mathcal{F}, M_k))}{M_k}} \right]
$$

$$
\leq \sum_{k=1}^{K} \frac{M_k \omega_k}{M} \sqrt{8 \left(\frac{\ln 2}{M_k} + \frac{\mathcal{V}}{M_k} \ln \frac{e M_k}{\mathcal{V}} \right)}
$$

Taking Janson's decomposition (4.11) with $Z_i = 1$ for all i, we see that $\sum_{k=1}^{K} M_k \omega_k = M$. From Jensen's inequality, we obtain:

$$
\mathcal{R}_m^{\mathfrak{T}}(L \circ \mathcal{F}) \leq \sqrt{8 \sum_{k=1}^{M} \left(\frac{M_k \omega_k \ln 2}{M M_k} + \frac{M_k \omega_k \mathcal{V}}{M M_k} \ln \frac{e M_k}{\mathcal{V}} \right)}
$$

$$
= \sqrt{8 \left(\frac{\chi^*(\mathfrak{T}) \ln 2}{M} + \frac{\mathcal{V}}{M} \sum_{k=1}^{K} \omega_k \ln \frac{e M_k}{\mathcal{V}} \right)}.
$$

From Jensen's inequality and the concavity of the logarithm, we have

$$
\frac{\mathcal{V}}{M} \sum_{k=1}^{K} \omega_k \ln \frac{e M_k}{\mathcal{V}} = \frac{\chi^*(\mathfrak{T}) \mathcal{V}}{M} \sum_{k=1}^{K} \frac{\omega_k}{\chi^*(\mathfrak{T})} \ln \frac{e M_k}{\mathcal{V}}
$$

$$
\leq \frac{\chi^*(\mathfrak{T}) \mathcal{V}}{M} \ln \frac{e \sum_{k=1}^{K} \frac{\omega_k M_k}{\chi^*(\mathfrak{T})}}{\mathcal{V}}
$$

The final result follows from $\sum_{k=1}^{K} \omega_k M_k = M$.

2. The second inequality follows from similar (but simpler) arguments using the bound on the empirical Rademacher complexity (2.33).

4.3.4   Application to Bipartite Ranking

In this section, we provide examples of explicit bounds that can be found using Theorems 4.3 and 4.4 on ranking tasks with (conditional) pairwise comparison risks. These are only examples, and many other bounds can be found with similar techniques.

We first consider error bounds based on the VC dimension. In the second part of this subsection, we present bounds for kernel machines.

The theorem below presents the fractional Rademacher bound for the AUC risk in bipartite ranking (Eq. (4.5) in Sect. 4.1.1). It is the the generalization error bound that extends the test set bound for bipartite ranking of Corollary 4.1.

Theorem 4.5 *Let $\mathcal{D}$ be a probability distribution over $\mathcal{X} \times \{-1, 1\}$, $\mathcal{F}$ be a class of functions from $\mathcal{X}$ to $\mathbb{R}$ and let $\mathcal{E}^{AUC}$ be the AUC risk defined by (4.5). Let furthermore $\tilde{\mathcal{F}}$ be the class of classifiers of pairs associated to $\mathcal{F}$:*

$$\tilde{\mathcal{F}} = \{(x, x') \in \mathcal{X}^2 \mapsto 2\mathbb{1}_{[f(x) > f(x')]} - 1 | f \in \mathcal{F}\}$$

and let us denote by $\mathcal{V}$ the VC dimension of $\tilde{\mathcal{F}}$.

Additionally, let $\mathcal{D}_{+1}$ and $\mathcal{D}_{-1}$ denote respectively the conditional distributions over $\mathcal{X} \times \{-1, 1\}$ of positive and negative examples, and let $(m^+, m^-) \in (\mathbb{N} \backslash \{0\})^2$.

If $m^+ \leq m^-$, then, for all $\delta \in (0, 1]$, with probability $1 - \delta$ over training set generated according to $\mathcal{D}_{+1}^{m^+} \otimes \mathcal{D}_{-1}^{m^-}$, the following bound holds for all $f \in \mathcal{F}$:

$$\mathcal{E}^{AUC}(f) \leq E^{AUC}(f, S) + \sqrt{8\left(\frac{\ln 2}{m^+} + \frac{\mathcal{V}}{m^+}\ln\frac{em^+}{\mathcal{V}}\right)} + 3\sqrt{\frac{\ln(2/\delta)}{2m^+}}.$$

If $m^+ \geq m^-$, the same bound, replacing m^+ by m^-, holds.

Generalization error bounds for the AUC were first proposed by Agarwal et al., (2005). In contrast to our bound, their proof uses a specific capacity measure for bipartite ranking that can be tighter than the VC dimension of the class of classifiers of pairs. However, their approach is specific to bipartite ranking, and cannot be applied to any other ranking scenarios.

More similarly to Theorem 4.5, Proposition 2 of Clémençon et al. (2005) gives a bound based on the VC dimension of $\tilde{\mathcal{F}}$ on the pairwise comparison risk for instance ranking. A bound for the AUC in the bipartite case can be derived from that latter bound if one knows the marginal probability of class 1. However, their approach, based on U-statistics, does not explicitly take into account the observed number of positive and negative examples. A bound similar to theirs could be obtained within our framework with arguments that are similar that those used for bipartite ranking. The main difference with the bipartite case is that we should consider the pairwise comparison risk (4.1) instead of the AUC risk; in the context of the pairwise comparison risk for ranking instances, we have $\frac{\chi^*(\mathfrak{I})}{M} = n$.

PROOF **of Theorem 4.5**

Let $\mathfrak{T}$ be the transformation function defined for bipartite ranking in (4.9) under the assumption $S \sim \mathcal{D}_{+1}^{m^+} \otimes \mathcal{D}_{-1}^{m^-}$. By construction, we know that $\mathcal{E}^{AUC}(f) = \mathcal{E}^{\mathfrak{T}}(f)$ and that $E^{AUC}(f, S) = E^{\mathfrak{T}}(f, S)$. We also know from Corollary 4.1 that $\frac{\chi^*(\mathfrak{T})}{M} = \frac{1}{\min(m^+, m^-)}$.

Now, denoting $S = ((x_1, y_1), ..., (x_{m^+ + m^-}, y_{m^+ + m^-}))$, we recall that a transformed example has the form $\tau_i = (1, (x_{i_1}, x_{i_2}), 1)$, where $i \in \{1, ..., m^+ m^-\}$, $i_1 \in \{1, ..., m^+\}$ and $i_2 \in \{m^+ + 1, ..., m^+ + m^-\}$. For any $\tilde{z} \in \{0, 1\}$, any $(x, x') \in \mathcal{X}^2$ and any $\tilde{y} \in \{-1, 1\}$, we then define the instantaneous loss function $L(\tilde{f}, (\tilde{z}, (x, x'), \tilde{y}) = \mathbb{1}_{[\tilde{f}(x, x') \neq 1]}$, where $\tilde{f}$ is the classifier of pairs associated to function $f \in \mathcal{F}$.

It can easily be seen that the VC dimension of $L \circ \tilde{\mathcal{F}}$ is equal to $\mathcal{V}$, and that $\mathcal{E}^{AUC}(f) = \mathcal{E}_L^{\mathfrak{T}}(\tilde{f})$ (the same holds for the empirical counterpart of these risks). Consequently, the first fractional Rademacher bound of Theorem 4.3 and the bound on the fractional Rademacher complexity using the VC dimension given by 4.4 directly lead to the desired result.

4.3.5 Application to Ranking of Alternatives for Multiclass Data

Taking into account the interdependency does not only allow to tackle problems of ranking of instances, but some cases of ranking of alternatives. We now describe how our approach can be applied to the case of label ranking for multiclass classification described in Sect. 4.1.2. From remark 4.2 the empirical risk for multiclass classification is defined by (4.6), which we repeat below:

$$\mathcal{E}^{mc}(f) = \mathbb{E}_{S \sim \mathcal{D}^m}[E^{mc}(f, S]$$

$$\text{with } E^{mc}(f, S) = \frac{1}{m(K-1)} \sum_{i=1}^{m} \sum_{k=2}^{K} \mathbb{1}_{[f(x_i^{\mu(y_i, 1)}) \leq f(x_i^{\mu(y_i, k)})]}.$$

in terms of notation, we remind that in our framework, the examples are i.i.d. pairs of sequences $(\mathbf{x} = (x^1, ..., x^K), \mathbf{y} = (y^1, ..., y^K))$, where K is the number of classes. The label vector $\mathbf{y}$ is a boolean vector with a single 1 at position $\mu(\mathbf{y}, 1)$, and the x^k are the joint representation of a class label and an input $x \in \mathcal{X} = \mathbb{R}^d$. Figure 4.4 (left) gives a representation of such data.

We also have that each of the $x^1, ..., x^K$ is described in a subset $\mathcal{X}'$ of $\mathcal{X}^K$ such that for x^k, only d components between indexes $\{k * d + 1, ..., (k + 1) * d\}$ are nonzero. Thus, if $\mathcal{F}$ is a class of scoring functions for the multiclass label ranking problem, then each f in '$ClsFncF$ is a function from $\mathcal{X}'$ to $\mathbb{R}$. We denote by $\tilde{\mathcal{F}}$ the class of classifiers of pairs induced by $\mathcal{F}$ using (4.7), and $\mathcal{V}$ the VC dimension of the

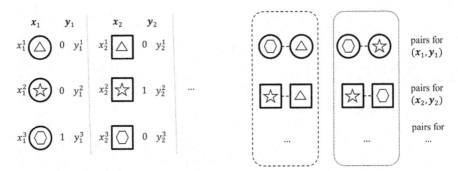

Fig. 4.4 *(left)* Illustration of the two first examples $(\mathbf{x}_1, \mathbf{y}_1)$ and $(\mathbf{x}_2, \mathbf{y}_2)$ *(circle* and *square)* for data labeled with one among $K = 3$ classes *(triangle, star, hexagon).* We have feature representation for each couple (example, class) represented by the small class symbol inside the bigger example symbol. Each of these couples has a label (1 or 0), but the examples belong to a single class: $\mathbf{x}_1$ *(circle)* belongs to the 3rd class *(hexagon)* and $\mathbf{x}_2$ *(square)* belongs to the 2nd class *(star). (right)* Illustration of a cover of size $K - 1$ of the pairs into independent sets. In the k-th subset of the cover, each original example appears in exactly one pair, ensuring that the pairs are independent

latter. Notice that contrarily to the case of ranking of instances, the scoring functions are not defined on the instance space (which here is $(\mathcal{X}')^K$). With such notation in mind, we can state the following theorem:

Theorem 4.6 *With the notation of the above paragraph, let $S \sim \mathcal{D}^m$ where $\mathcal{D}$ is a distribution over data classified in exactly one out of K classes. Then, for all $\delta \in (0, 1]$, with probability at least $1 - \delta$, the following bound holds for all scoring functions $f \in \mathcal{F}$:*

$$\mathcal{E}^{mc}(f) \leq E^{mc}(f, S) + \sqrt{8\left(\frac{\ln 2}{m} + \frac{V}{m}\ln\frac{em}{V}\right)} + 3\sqrt{\frac{\ln(2/\delta)}{2m}}.$$

PROOF **of Theorem 4.6**

The proof uses the transformation for multiclass data $\mathfrak{T}$ defined in (4.10) of Sect. 4.2.1. Using this transformation, we can directly apply of the bound based on the fractional Rademacher complexity of Theorem 4.3 and its bound using the VC dimension of Theorem 4.4, we only get to compute $\frac{\chi^*(\mathfrak{T})}{M}$ corresponding to $\mathfrak{T}$.

In that case, $M = m(K - 1)$, and it is easy that we can build a (fractional) cover $((ccover_k, \mathfrak{M}_k))_{k=1}^{K-1}$ of size $K - 1$ using independent sets of $\mathfrak{T}(S)$ as described in Fig. 4.4, and which can be formalized as follows:

$$\forall k \in \{1, ..., K - 1\}, \omega_k = 1 \text{ and:}$$

$$\mathfrak{M}_k = ((1, (x_1^{\mu(\mathbf{y}_i, 1)}, x_1^{\mu(\mathbf{y}_i, k+1)}), 1),$$

$$(1, (x_2^{\mu(\mathbf{y}_i,1)}, x_2^{\mu(\mathbf{y}_i,k+1)}), 1)$$

$$\dots$$

$$1, (x_m^{\mu(\mathbf{y}_i,1)}, x_m^{\mu(\mathbf{y}_i,k+1)}), 1)).$$

It is clear the transformed examples in each $\mathfrak{M}_k$ are independent since each pair considers a different instance. We obtain $\chi^*(\mathfrak{T}) \le (K-1)$, from which we conclude $\frac{\chi^*(\mathfrak{T})}{M} = \frac{1}{m}$.

References

Agarwal, S. (2012). Surrogate regret bounds for bipartite ranking via strongly proper losses. *Journal of Machine Learning Research, 15*, 1653–1674. arXiv preprint arXiv:1207.0268.

Agarwal, S., Graepel, T., Herbrich, R., Har-Peled, S., & Roth, D. (2005). Generalization bounds for the area under the roc curve. *Journal of Machine Learning Research, 6*, 393–425.

Amini, M.-R., Usunier, N., & Goutte, C. (2010). Learning from multiple partially observed views - an application to multilingual text categorization. In *Advances in Neural Information Processing Systems (NIPS 22)*, pages 28–36.

Amini, M.-R., Usunier, N., & Laviolette, F. (2009). A transductive bound for the voted classifier with an application to semi-supervised learning. In *Advances in Neural Information Processing Systems (NIPS 21)*, pages 65–72.

Antos, A., Kégl, B., Linder, T., & Lugosi, G. (2003). Data-dependent margin-based generalization bounds for classification. *Journal of Machine Learning Research, 3*, 73–98.

Bach, F., Lanckriet, R., & Jordan, M. (2004). Multiple kernel learning, conic duality, and the smo algorithm. In *Proceedings of the Twenty-first International Conference on Machine Learning*.

Bartlett, P. L., & Mendelson, S. (2003a). Rademacher and gaussian complexities: Risk bounds and structural results. *Journal of Machine Learning Research, 3*, 463–482.

Bartlett, P. L., & Mendelson, S. (2003b). Rademacher and gaussian complexities: Risk bounds and structural results. *Journal of Machine Learning Research, 3*, 463–482.

Basu, S., Banerjee, A., & Mooney, R. J. (2002). Semi-supervised clustering by seeding. In *Proceedings of the Nineteenth International Conference on Machine Learning*, pages 27–34.

Blitzer, J., Crammer, K., Kulesza, A., Pereira, F., & Wortman, J. (2007). Learning bounds for domain adaptation. In *NIPS*.

Blum, A., & Mitchell, T. (1998a). Combining labeled and unlabeled data with co-training. In *Proceedings of the 11th Annual Conference on Learning Theory*, pages 92–100.

Blum, A., & Mitchell, T. M. (1998b). Combining labeled and unlabeled sata with co-training. In *COLT*, pages 92–100.

Blumer, A., Ehrenfeucht, A., Haussler, D., & Warmuth, M. (1989). Learnability and the vapnik-chervonenkis dimension. *Journal of the ACM, 36*, 929–965.

Boucheron, S., Bousquet, O., & Lugosi, G. (2005). Theory of classification: A survey of some recent advances. *ESAIM: Probability and Statistics*, pages 323–375.

Bronnimann, H., & Goodrich, M. T. (1995). Almost optimal set covers in finite vc-dimension. *Discrete and Computational Geometry, 14*(4), 463–479.

Burges, C. J. (2010). From ranknet to lambdarank to lambdamart: An overview. *Learning, 11*, 23–581.

Calauzènes, C., Usunier, N., & Gallinari, P. (2013). Calibration and regret bounds for order-preserving surrogate losses in learning to rank. *Machine Learning, 93*(2–3), 227–260.

© Springer International Publishing Switzerland 2015

M.-R. Amini, N. Usunier, *Learning with Partially Labeled and Interdependent Data*,

DOI 10.1007/978-3-319-15726-9

Cao, Z., Qin, T., Liu, T.-Y., Tsai, M.-F., & Li, H. (2007). Learning to rank: From pairwise approach to listwise approach. In *Proceedings of the 24th international conference on Machine learning*, pages 129–136. ACM.

Celeux, G., & Govaert, G. (1992). A classification em algorithm for clustering and two stochastic versions. *Computational Statistics and Data Analysis, 14*(3), 315–332.

Cesa-Bianchi, N., & Haussler, D. (1998). A graph-theoretic generalization of the sauer-shelah lemma. *Discrete Applied Mathematics, 86,* 27–35.

Chapelle, O., & Chang, Y. (2011). Yahoo! learning to rank challenge overview. In *Yahoo! Learning to Rank Challenge*, pages 1–24.

Chapelle, O., Schölkopf, B., & Zien, A. (Eds.). (2006). *Semi-supervised learning.* Cambridge: MIT Press.

Clémençon, S., Lugosi, G., & Vayatis, N. (2005). Ranking and scoring using empirical risk minimization. In *Learning Theory*, pages 1–15. Springer.

Clémençon, S., Robbiano, S., & Vayatis, N. (2013). Ranking data with ordinal labels: Optimality and pairwise aggregation. *Machine Learning, 91*(1), 67–104.

Clémençon, S., & Vayatis, N. (2007). Ranking the best instances. *The Journal of Machine Learning Research, 8,* 2671–2699.

Clémençon, S., & Vayatis, N. (2009). Tree-based ranking methods. *Information Theory, IEEE Transactions on, 55*(9), 4316–4336.

Clémençon, S., & Vayatis, N. (2010). Overlaying classifiers: A practical approach to optimal scoring. *Constructive Approximation, 32*(3), 619–648.

Cohen, I., Cozman, F. G., Sebe, N., Cirelo, M. C., & Huang, T. S. (2004). Semisupervised learning of classifiers: Theory, algorithms, and their application to human-computer interaction. *IEEE Transactions on Pattern Analysis and Machine Intelligence, 26*(12), 1553–1567.

Cohen, W. W., Schapire, R. E., & Singer, Y. (1998). Learning to order things. In *Advances in Neural Information Processing Systems (NIPS 10)*, pages 451–457.

Cortes, C., & Mohri, M. (2004). *AUC optimization vs. error rate minimization. In Advances in Neural Information Processing Systems (NIPS 16)*, pages 313–320.

Cossock, D., & Zhang, T. (2008). Statistical analysis of bayes optimal subset ranking. *Information Theory, IEEE Transactions on, 54*(11), 5140–5154.

Crammer, K., & Singer, Y. (2002). Pranking with ranking. In *Advances in Neural Information Processing Systems (NIPS 14)*, pages 641–647.

Dantzig, G. (1951). Maximization of a linear function of variables subject to linear inequalities. In T. Koopmans (Ed.), *Activity analysis of production and allocation* (pp. 339–347). New York: Wiley.

Dekel, O., Manning, C., & Singer, Y. (2003). Log-linear models for label ranking.

Dempster, A. P., Laird, N. M., & Rubin, D. B. (1977). Maximum likelihood from incomplete data via the em algorithm. *Journal of the Royal Statistical Society. Series B (Methodological), 39*(1), 1–38.

Derbeko, P., El-Yaniv, E., & Meir, R. (2003). Error bounds for transductive learning via compression and clustering. In *Advances in Neural Information Processing Systems (NIPS 15)*, pages 1085–1092.

Duchi, J. C., Mackey, L., Jordan, M. I., et al. (2013). The asymptotics of ranking algorithms. *The Annals of Statistics, 41*(5), 2292–2323.

Duda, R., Hart, P., & Stork, D. (2001). *Pattern classification.* Wiley.

Ehrenfeucht, A., Haussler, D., Kearns, M., & Valiant, L. (1989). A general lower bound on the number of examples needed for learning. *Information and Computation, 82,* 247–261.

Fralick, S. C. (1967). Learning to recognize patterns without a teacher. *IEEE Transactions on Information Theory, 13*(1), 57–64.

Freund, Y., Iyer, R., Schapire, R. E., & Singer, Y. (2003). An efficient boosting algorithm for combining preferences. *Journal of Machine Learning Research, 4,* 933–969.

Fukunaga, K. (1972). *Introduction to statistical pattern recognition.* New York: Academic Press.

Genesereth, M. R., & Nilsson, N. J. (1987). *Logical foundations of artificial intelligence*. San Francisco: Morgan Kaufmann Publishers Inc..

Giné, E. (1996). Empirical processes and applications: An overview. *Bernoulli, 2*(1), 1–28.

Grandvalet, Y., & Bengio, Y. (2005). Semi-supervised learning by entropy minimization. In *Advances in Neural Information Processing Systems (NIPS 17)*, pages 529–536. MIT Press.

Hand, D. J., & Till, R. J. (2001). A simple generalisation of the area under the roc curve for multiple class classification problems. *Machine Learning, 45*(2), 171–186.

Har-Peled, S., Roth, D., & Zimak, D. (2002).Constraint classification: A new approach to multiclass classification. In *Algorithmic Learning Theory*, pages 365–379. Springer.

Herbrich, R., Graepel, T., Bollmann-Sdorra, P., & Obermayer, K. (1998). Learning preference relations for information retrieval. In *Proceedings of the AAAI Workshop Text Categorization and Machine Learning, Madison, USA*.

Hoeffding, W. (1963). Probability inequalities for sums of bounded random variables. *Journal of the American Statistical Association, 58*, 13–30.

Janson, S. (2004). Large deviations for sums of partly dependent random variables. *Random Structures and Algorithms, 24*(3), 234–248.

Järvelin, K., & Kekäläinen, J. (2002). Cumulated gain-based evaluation of ir techniques. *ACM Transactions on Information Systems (TOIS), 20*(4), 422–446.

Joachims, T. (2002a). *Learning to classify text using support vector machines: Methods, theory and algorithms*. Norwell: Kluwer Academic Publishers.

Joachims, T. (2002b). Optimizing search engines using clickthrough data. In *Proceedings of the eighth ACM SIGKDD international conference on Knowledge discovery and data mining*, pages 133–142. ACM.

Joachims, T. (2005). A support vector method for multivariate performance measures. In *Proceedings of the 22nd international conference on Machine learning*, pages 377–384. ACM.

Joachmis, T. (1999). Transductive inference for text classification using support vector machines. In *Proceedings of the 16th International Conference on Machine Learning*, pages 200–209.

Koltchinskii, V. I. (2001). Rademacher penalties and structural risk minimization. *IEEE Transactions on Information Theory*, 47(5):1902–1914.

Koltchinskii, V. I., & Panchenko, D. (2000). Rademacher processes and bounding the risk of function learning. In E. Gine, D. Mason, & J. Wellner (Eds.), *High dimensional probability II*, pp. 443–459.

Kotlowski, W., Dembczynski, K. J., & Huellermeier, E. (2011). Bipartite ranking through minimization of univariate loss. In *Proceedings of the 28th International Conference on Machine Learning (ICML-11)*, pages 1113–1120.

Langford, J. (2005). Tutorial on practical prediction theory for classification. *Journal of Machine Learning Research, 6*, 273–306.

Langford, J., & Shawe-Taylor, J. (2002). Pac-bayes & margins. *NIPS, 15*, 439–446.

Latouche, G., & Ramaswami, V. (1999). *Introduction to matrix analytic methods in stochastic modeling*. ASA-SIAM Series on Statistics and Applied Probability. Philadelphia, Pa. SIAM, Society for Industrial and Applied Mathematics Alexandria, Va. ASA, American Statistical Association.

Ledoux, M., & Talagrand, M. (1991). *Probability in Banach spaces: Isoperimetry and processes*. Springer Verlag.

Leskes, B. (2005a). The value of agreement, a new boosting algorithm. In *Proceedings of Conference on Learning Theory (COLT)*, pages 95–110.

Leskes, B. (2005b). The value of agreement, a new boosting algorithm. *Lecture Notes in Computer Science, 3559*, 95–110. *COLT*

Liu, T.-Y. (2009). Learning to rank for information retrieval. *Found Trends Inf Retr, 3*(3), 225–331.

Lugosi, G. (2002). Pattern classification and learning theory. In L. Gyorfi, (Eds), *Principles of nonparametric learning*, 1–56. Springer.

Machlachlan, G. J. (1992). *Discriminant analysis and statistical pattern recognition*. Wiley Interscience.

Manning, C. D., Raghavan, P., & Schütze, H. (2008). *Introduction to information retrieval*. New York: Cambridge University Press.

Massart, P. (2000). Some applications of concentration inequalities to statistics. *Annales de la faculté des sciences de Toulouse, 9*(2), 245–303.

McAllester, D. A. (2003). Pac-bayesian stochastic model selection. *Machine Learning, 51*(1), 5–21.

McDiarmid, C. (1989). On the method of bounded differences. *Surveys in combinatorics, 141,* 148–188.

Mohri, M., Rostamizadeh, A., & Talwalkar, A. (2012). *Foundations of machine learning*. MIT Press.

Montroll, E. (1956). Random walks in multidimensional spaces, especially on periodic lattices. *Journal of the Society for Industrial and Applied Mathematics (SIAM), 4*(4), 241–260.

Muslea, I. (2002). *Active learning with multiple views*. PhD thesis, USC.

Nigam, K., McCallum, A. K., Thrun, S., & Mitchell, T. (2000). Text classification from labeled and unlabeled documents using EM. *Machine Learning Journal, 39*(2–3), 103–134.

Patrick, E. A., Costello, J. P., & Monds, F. C. (1970). Decision-directed estimation of a two-class decision boundary. *IEEE Transactions on Information Theory, 9*(3), 197–205.

Quoc, C., & Le, V. (2007). Learning to rank with nonsmooth cost functions. *NIPS07, 19,* 193.

Rajaram, S. & Agarwal, S. (2005). Generalization bounds for k-partite ranking. In *Proceedings of the NIPS workshop on learning to rank*, pages 18–23.

Ravikumar, P. D., Tewari, A., & Yang, E. (2011). On ndcg consistency of listwise ranking methods. In *International Conference on Artificial Intelligence and Statistics*, pages 618–626.

Rudin, C. (2009). The p-norm push: A simple convex ranking algorithm that concentrates at the top of the list. *The Journal of Machine Learning Research, 10,* 2233–2271.

Rudin, C., Cortes, C., Mohri, M., & Schapire, R. E. (2005). Margin-based ranking meets boosting in the middle. In *Conference On Learning Theory (COLT)*.

Sauer, N. (1972). On the density of families of sets. *Journal of Combinatorial Theory, 13*(1), 145–147.

Seeger, M. (2001). Learning with labeled and unlabeled data. Technical report.

Shelah, S. (1972). A combinatorial problem: Stability and order for models and theories in infinity languages. *Pacific Journal of Mathematics, 41,* 247–261.

Sindhwani, V., Niyogi, P., & Belkin, M. (2005). A co-regularization approach to semi-supervised learning with multiple views. In *ICML-05 Workshop on Learning with Multiple Views*, pages 74–79.

Szummer, M., & Jaakkola, T. (2002). Partially labeled classification with markov random walks. In *Advances in Neural Information Processing Systems (NIPS 14)*, pages 945–952.

Taylor, J., & Cristianini, N. (2004). *Kernel methods for pattern analysis*. New York: Cambridge Press University.

Tchebychev, P. L. (1867). Des valeurs moyennes. *Journal de mathématiques pures et appliquées, 2*(12), 177–184.

Tür, G., Hakkani-Tür, D. Z., & Schapire, R. E. (2005). Combining active and semi-supervised learning for spoken language understanding. *Speech Communication, 45*(2), 171–186.

Usunier, N., Amini, M.-R., & Gallinari, P. (2006). Generalization error bounds for classifiers trained with interdependent data. In *Advances in Neural Information Processing Systems (NIPS 18)*, pages 1369–1376.

Vapnik, V. N. (1999). *The nature of statistical learning theory* (2nd ed.). New York: Springer-Verlag.

Vapnik, V. N., & Cervonenkis, A. J. (1971). On the uniform convergence of relative frequencies of events to their probabilities. *Theory of Probability and its Applications, 16,* 264–280.

Vapnik, V. N., & Cervonenkis, A. J. (1974). *Theory of pattern recognition*. Moscow: Nauka.

Voorhees, E. M., Harman, D. K., et al. (2005). *TREC: Experiment and evaluation in information retrieval* (Vol 63). Cambridge: MIT press.

Weston, J. & Watkins, C. (1999). Support vector machines for multi-class pattern recognition. In *European Symposium on Artificial Neural Netwroks (ESANN)*, pages 219–224.

Zheng, Z., Zha, H., Zhang, T., Chapelle, O., Chen, K., & Sun, G. (2008). A general boosting method and its application to learning ranking functions for web search. In *Advances in neural information processing systems*, pages 1697–1704.

Zhou, D., Bousquet, O., Lal, T. N., Weston, J., & Schölkopf, B. (2004). Learning with local and global consistency. In *Advances in Neural Information Processing Systems (NIPS 16)*, pages 321–328. MIT Press.

Zhu, X. & Ghahramani, Z. (2002). Learning from labeled and unlabeled data with label propagation. Technical report CMU-CALD-02-107, Carnegie Mellon University.

Zhu, X., Ghahramani, Z., & Lafferty, J. (2003). Semi-supervised learning using gaussian fields and harmonic functions. In *20th International Conference on Machine Learning*, pages 912–919.

Index

© Springer International Publishing Switzerland 2015
M.-R. Amini, N. Usunier, *Learning with Partially Labeled and Interdependent Data,*
DOI 10.1007/978-3-319-15726-9

Printed in the United States
By Bookmasters